Newnes Construction Materials Pocket Book

Newnes Construction Materials Pocket Book

D.K. DORAN
BSc (Eng), DIC, FCGI, FICE, FIStructE
*Consultant, formerly Chief Structural Engineer
George Wimpey plc*

 NEWNES

Newnes
An imprint of Butterworth–Heinemann Ltd
Linacre House, Jordan Hill, Oxford OX2 8DP

 A member of the Reed Elsevier group plc

OXFORD   LONDON   BOSTON
MUNICH   NEW DELHI   SINGAPORE   SYDNEY
TOKYO   TORONTO   WELLINGTON

First published 1994

British Library Cataloguing in Publication Data

Doran, David K.
  Newnes Construction Materials Pocket Book
  I. title
  624.18

ISBN 0 7506 1666 0

Library of Congress Cataloguing in Publication Data

Doran, David K.
  Newnes construction materials pocket book/D. K. Doran
  p.   cm.
  Includes bibliographical references and index.
  ISBN 0 7506 1666 0
  1. Building materials–Handbooks, manuals, etc. I. title.
  II. Title: Construction materials pocket book.
  TA403.4.D67
  624.1′82–dc20   94–26230
                                                         CIP

Typeset by TecSet Ltd, Wallington, Surrey
Printed in Great Britain by Clays Ltd, St Ives plc

Contents

Preface

The number of materials used in the construction process is too numerous to estimate. We have tried, therefore, to identify those most widely used. Some, such as metals, are used in elemental form; others combine to provide composites of great strength, stiffness and durability.

This pocket book, esssentially a digest of up-dated basic data from the *Construction Materials Reference Book* (Butterworth–Heinemann 1992) has been produced to meet the requirements of student and practitioner alike. The book is intended as a speedy reference for busy people and a guide to further information if required.

I am greatly indebted to the authors of the *Construction Materials Reference Book* without whose expertise this book would not have been possible. I trust that my distillation of their work meets with their approval.

My thanks are due to Paddy Baker who first suggested the project, to Neil Warnock-Smith my Commissioning Editor and to my wife, Maureen, for her encouragement and help in the many tasks that go to produce a book such as this.

David Doran
Wanstead

Notation

Units of measurement in this book are given using the system of negative indices favoured in the SI system, rather than oblique strokes. Under this system, a unit of measurement which would appear *after* an oblique stroke (or in the *denominator* of a fraction) is given a negative index.

For example:

- newtons per square millimetre is written $N\,mm^{-2}$ (rather than N/mm^2)
- watts per square metre per degree Celsius is written $W\,m^{-2}\,°C^{-1}$ (rather than $W/m^2/°C$ or $W/m^2°C$)

Note that this only applies to a unit which is being *divided into* another (metres, square millimetres, square metres or degrees Celsius in the previous examples). Negative indices are not used in the following examples:

- *Torque* has dimensions of force × distance, and so can be measured in newtons × metres, or newton metres ($N\,m$)
- *Electrical energy* has dimensions of power × time, and so can be measured in kilowatts × hours, or kilowatt hours ($kW\,h$).

For further information, the reader is referred to *Quantities, Units and Symbols*, published by The Symbols Committee of The Royal Society, The Royal Society, London, 1975.

Where conversion to Imperial units required please consult the CIRIA publication *Metric Conversion Factors*.

Contributors

W J Allwood BSc
BP Chemicals

R H Andrews
Pemtech Associates Ltd

G H Arnold BSc, CEng, MIChemE
European Vinyls Corporation (UK) Ltd

D C Aslin BA
Prometheus Developments Ltd

S Austin BSc, PhD, CEng, MICE
Loughborough University of Technology

R C Baker BSc, CEng, MICE
Grove Consultants Ltd

M J Bayley MSc(Eng)
Technical Support Manager, Alcan Offshore

C E G Bland FIWEM, FRSH
Clay Pipe Development Association Ltd

C Bodsworth MMet, PhD, CEng, FIM
Brunel, The University of West London

R G Bristow BSc(Eng), CEng, MICE, MIStructE
G Maunsell and Partners

P S Bulson CBE, DSc, FEng
Chairman, British Standards Committee on the Structural Use
of Aluminium

C J Burgoyne PhD
Engineering Department, Cambridge University

R N Butlin PhD
Building Research Establishment

F G Buttler PhD, MRSC, CChem
Consultant Chemist

N C Clark
Mandoval Ltd

V A Coveney PhD
Bristol Polytechnic

J Crisfield HND(BLDG)
Vencel Resil Ltd

J V Crookes MSc, CChem, MRSC
Cuprinol Ltd

D A Cross BSc(Hons), DMS
Vencel Resil Ltd

S H Cross CPhys, MInstP, AMI, CorrST
The Glassfibre Reinforced Cement Association

D E J Cunningham MICorrST
Herbert Industrial Coatings

R Dennis CChem, MRSC
Doverstrand Ltd

J Dodd AIoR
The Marley Roof Tile Co Ltd

DK Doran BSc Eng, CEng, DIC, FCGI, FICE, FIStructE
Consultant

J W Dougill MSc(Eng), DIC, PhD, FEng, FICE, FIStructE,
FASCE
Institution of Structural Engineers

J J Farrington CEng, MICE
Staffordshire County Council

M G Grantham BA, MRSC, MIQA, MIHT, CChem
Technotrade Ltd

R Harris BSc
Wimpey Environmental Ltd

B A Haseltine BSc(Eng), FCGI, DIC, FEng, FICE, FIStructE,
FICeram, MConsE
Jenkins & Potter Consulting Engineers

L Hodgkinson
Cormix Construction Chemicals
Chairman, Technical Committee, Cement Admixtures
Association

S A Hurley BSc, PhD
Taywood Engineering Ltd

M L Humpage BTech, CChem, MRSC
Fosroc Technology

T S Ingold BSc, MSc, PhD, DIC, EurIng, CEng, MConsE, FICE, FIHT, FGS, FASCE, MSocIS (France)
Consulting Geotechnical Engineer

G K Jackson
Pilkington Glass Consultants

S A Jefferis MA, PhD, CEng, MICE, FGS
W J Engineering Resources Ltd

M Kawamura Dr Eng
Kanazawa University, Japan

R I Lancaster FICE, CEng, FIStructE, FCIArb, FACI
Consulting Engineer

C D Lawrence
British Cement Association

M Malinowsky
Consulting Engineer

J R Moon PhD
University of Nottingham

K Nakano Dr Eng
Osaka Cement Co. Ltd, Japan

P Olley
C Olley & Sons Ltd

D O'Sullivan
Formerly of British Gypsum plc

J Pitts PhD, BSc, DipEng, CEng, MIMM, MIGeol, MIQA, FGS
Geotechnical and Environmental Consultants Ltd

C D Pomeroy DSc, CPhys, FInstP, FACI, FSS
Fomerly of the British Cement Association

B Ralph MA, PhD, ScD, CEng, CPhys, FIM, FInstP, Hon.FRMS, Eur.Ing
Brunel, The University of West London

P J Ridd BSc(Eng), DMS
Cemfil International Ltd

P Robins BSc, PhD, CEng, MICE
Loughborough University of Technology

K D Ross
Building Research Establishment

J D N Shaw
SBD Ltd

P M Smith
Eurosil: UK Mineral Wool Association

J G Sunley MSc, CEng, FIStructE
Kewstoke Ltd

R J M Sutherland BA, FEng, FICE, FIStructE
Consultant to Harris & Sutherland

R A Sykes BSc
Wimpey Environmental Ltd

L J Tabor FPRI
Fosroc CCD Ltd

D J Thompsett MA(Cantab)
Vencel Resil Ltd

A K Tovey CEng, FIStructE, ACIArb, MSFSE
Building and Structures Department, British Cement
Association

Note: Affiliations are those that applied at time of publication
of the *Construction Materials Reference Book*.

Part One: Metals

1 *Aluminium*

1.1 Introduction

Over the last century, aluminium has grown from a newly produced metal, first going into commercial production in 1886, to the second most used metal in constructional engineering. Aluminium alloys, each in a range of available forms, tempers and properties, have been developed to meet the needs of all forms of construction. The metal is light with good to excellent corrosion resistance and can be produced economically in a wide range of forms. The ease of extrusion of many of its alloys enables their use in complex sections in many applications throughout the construction industry. Successful use of the correct aluminium or aluminium alloy in all applications is founded on a sound understanding of the properties and other aspects of the material.

Aluminium, by reason of its chemical reactivity, is not found in its pure state, but in combination with other elements. It is the most common metallic element in the earth's crust which contains approximately 15% of Al_2O_3 (8% aluminium). In most locations the aluminium content is too small for economic extraction, but the principal ore of commercial value, and that most frequently used in the production of aluminium, is a group termed 'bauxites'. This contains mainly hydrated alumina together with oxides of iron, silicone and titanium. Bauxites are regarded as the end-products of the slow weathering of aluminium-bearing rocks, often igneous but not necessarily so, over millions of years in regions of heavy rainfall common in tropical or subtropical areas. Some bauxites occur as sedimentary or alluvial deposits from water which has carried them away from the weathering rocks. In either case the products are mainly surface deposits and are extracted by open-cast quarrying. The bauxite is then washed and screened to remove extraneous dirt.

The process of refining bauxite to alumina, indicating the material and energy requirements, is illustrated in *Figure 1.1*.

Aluminium is then smelted into ingots which may vary from 2 to 20 000 kg. The cast metal thus produced is then fabricated by various processes into forms of aluminium alloy suitable for the manufacture of finished products. These processes include:

- Casting (e.g. sand casting, gravity die casting, pressure die casting and other methods of casting).
- Rolling
- Extrusion (see examples in *Figures 1.2, 1.3* and *1.4*)
- Tube (e.g. seam-welded tube, extruded tube, drawn tube)

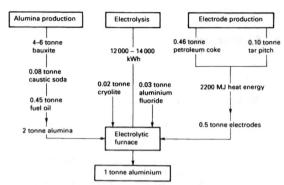

Figure 1.1   *Raw materials and energy requirements for the production of aluminium*

- Forging (e.g. die forging or hand forging)
- Superplastic forming.

(1) *Solid-shaped sections* are those which have no areas either wholly or partly enclosed (see *Figure 1.2*). These may be simple sections or complex sections with many special features.

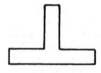

Figure 1.2   *Simple solid section*

(2) *Hollow-shaped sections* have one or more wholly enclosed areas (see *Figure 1.3*). These again may be simple hollow sections, such as round or rectangular tube or more complex sections involving additional features or multihollows.

Figure 1.3   *Simple hollow section*

(3) *Semi-hollow-shaped sections* are those which have partly enclosed features, having a substantial closed area in relation to the width of the gap (see *Figure 1.4*).

Figure 1.4   *Semi-hollow section*

1.2 Alloy types

Aluminium is seldom used in its pure form but is normally
alloyed with small proportions of other elements. Typical of
these are:

- Copper
- Manganese
- Silicon
- Magnesium
- Zinc.

There is no universal standard classification of aluminium
alloys although they are usually classified with respect to fabrication process, chemical composition and their ability to
respond to heat treatment or work hardening. This broad classification is indicated in *Figure 1.5*.

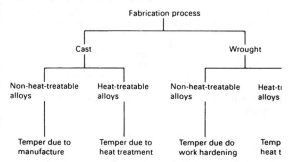

Figure 1.5 *Alloy classification*

Aluminium alloys are usually classified by a numerical designation in America or an alphanumeric designation in Europe.
Either system of identifying the chemical composition is
usually followed by a symbol or symbols which identify the
manufacturing process and temper. Thus in the UK, casting
alloys to BS1490 would be designated by numbers in sequence
of development which would be preceded by the letters LM
(light metal)—for example LM6. In North America the
Aluminium Association uses a four-digit system. The first
digit indicates the alloy group (see *Table 1.1*), the second and
third digits identify the aluminium alloy or purity and the last
digit, which is separated by a decimal point, indicates the product form (i.e. 0 indicates castings, 1 indicates ingots).

Examples of the two systems are shown in *Table 1.2*.

Table 1.1 *Major alloying elements in the four-digit classification system for cast alloys*

First digit	Major alloying element
1**.*	Pure aluminium (greater than 99%)
2**.*	Copper
3**.*	Silicon plus copper and/or magnesium
4**.*	Silicon
5**.*	Magnesium
6**.*	Unused series
7**.*	Zinc
8**.*	Tin
9**.*	Other elements

Table 1.2 *Examples of the four-digit numerical alloy nomenclature system*

Four-digit numerical system	Alphanumerical system	Major alloying elements
3103	Al Mn1	Manganese
5083	Al Mg4.5 Mn	Magnesium
6082	Al Si1 Mg Mn	Magnesium and silicon

1.3 Chemical composition and mechanical properties

The nominal composition of all alloys is defined in British or other national standards covering the wrought or cast forms and products. The composition of the major alloys in use in construction is shown in *Table 1.3*.

Material and product specifications normally define minimal values of mechanical properties. The three principal properties are:

- Minimum 0.2% proof stress
- Minimum tensile strength
- Elongation %.

These, together with other data, are presented in *Tables 1.4 to 1.8* for the following product forms:

1.4 Plate sheet and strip
1.5 Extruded bar, round tube and sections
1.6 Drawn tube and seam-welded tube
1.7 Ingots and castings
1.8 Forging stock and forgings

It should be noted that these properties will, in general, be dependent on the product form, thickness and temper or heat treatment of the metal. Aluminium in its pure state does not

Table 1.3 Chemical composition (%) of wrought and cast aluminium alloys[a]

Material designation	Silicon	Iron	Copper	Manganese	Magnesium	Chromium	Zinc	Other restrictions	Titanium	Others Each	Others Total	Aluminium
1080A	0.15	0.15	0.03	0.02	0.02	–	0.06	(Ga 0.03)	0.02	0.02	–	99.80 min
1050A	0.25	0.40	0.05	0.05	0.05	–	0.07	–	0.05	0.03	–	99.50 min
1200	1.0 Si + Fe		0.05	0.05	–	–	0.10	–	0.05	0.05	0.15	99.00 min
1350	0.10	0.40	0.05	0.01	–	0.01	0.05	(Ga 0.03, B 0.05, V + Ti 0.02)	–	0.03	0.10	99.50 min
3103	0.50	0.7	0.10	0.9–1.5	0.30	0.10	0.20	(Zr + Ti 0.10)	–	0.05	0.15	Rem
3105	0.6	0.7	0.30	0.30–0.8	0.20–0.8	0.20	0.40	–	0.10	0.05	0.15	Rem
5083	0.40	0.40	0.10	0.40–1.0	4.0–4.9	0.05–0.25	0.25	–	0.15	0.05	0.15	Rem
5154A	0.50	0.50	0.10	0.50	3.1–3.9	0.25	0.20	–	0.20	0.05	0.15	Rem
5251	0.40	0.50	0.15	0.10–0.50	1.7–2.4	0.15	0.15	(Mn + Cr 0.10–0.50)	0.15	0.05	0.15	Rem
5454	0.25	0.40	0.10	0.50–1.0	2.4–3.0	0.05–0.20	0.25	–	0.20	0.05	0.15	Rem
6061	0.40–0.8	0.7	0.15–0.40	0.15	0.8–1.2	0.04–0.35	0.25	–	0.15	0.05	0.15	Rem
6063	0.2–0.6	0.35	0.10	0.10	0.45–0.9	0.10	0.10	–	0.10	0.05	0.15	Rem
6063A	0.30–0.6	0.15–0.35	0.10	0.15	0.6–0.9	0.05	0.15	–	0.10	0.05	0.15	Rem
6082	0.7–1.3	0.50	0.10	0.40–1.0	0.6–1.2	0.25	0.20	–	0.10	0.05	0.15	Rem
7020	0.35	0.40	0.20	0.05–0.50	1.0–1.4	0.10–0.35	4.0–5.0	(Zr 0.08–0.20; Zr + Ti 0.08–0.25)	–	0.05	0.15	Rem
LM2	9.0–11.5	1.0	0.7–2.5	0.5	0.30	–	2.0	(Ni 0.5; Pb 0.3; Sn 0.2)	0.2	–	0.05	Rem
LM4	4.0–6.0	0.8	2.0–4.0	0.2–0.6	0.20	–	0.5	(Ni 0.3; Pb 0.1; Sn 0.1)	0.2	0.05	0.15	Rem
LM5	0.3	0.6	0.1	0.3–0.7	3.0–6.0	–	0.1	(Ni 0.1; Pb 0.05; Sn 0.05)	0.2	0.05	0.15	Rem
LM6	10.0–13.0	0.6	0.1	0.5	0.10	–	0.1	(Ni 0.1; Pb 0.1; Sn 0.05)	0.2	0.05	0.15	Rem
LM25	6.5–7.5	0.5	0.20	0.3	0.20–0.6	–	0.1	(Ni 0.1; Pb 0.1; Sn 0.05)	0.2	0.05	0.15	Rem

[a] Compositions are maximum values unless shown as a range or a minimum. Rem, remainder. Table partly compiled from BS 1470, BS 1474, BS 4300/17, BS 1490 and BS 2897.

Table 1.4 Mechanical properties of aluminium and aluminium alloy plate, sheet and strip materials[a]

Alloy designation British Standard	Alloy designation ISO standards	Product forms	Temper designation	Thickness (mm) Over	Thickness (mm) Up to and including	Minimum 0.2% proof stress (N mm⁻²)	Minimum tensile strength (N mm⁻²)	Elongation (%) 5.65√S₀	Elongation (%) 50 mm	Durability rating	Related British Standard
1080A	Al 99.80	Plate, sheet, strip	F	3.0	25.0	–	–	–	–	A	BS 1470
			O	0.2	6.0	–	90	–	29–35		
			H14	0.2	12.5	–	90	–	5–8		
			H18	0.2	3.0	–	125	–	3–5		
1050A	Al 99.50	Plate, sheet, strip	F	3.0	25.0	–	–	–	–	A	BS 1470
			O	0.2	6.0	–	55	–	22–32		
			H12	0.2	6.0	–	80	–	4–9		
			H14	0.2	12.5	–	100	–	4–8		
			H18	0.2	3.0	–	135	–	3–4		
1200	Al 99.0	Plate, sheet, strip	F	3.0	25.0	–	–	–	–	A	BS 1470
			O	0.2	6.0	–	70	–	20–30		
			H12	0.2	6.0	–	90	–	4–9		
			H14	0.2	12.5	–	105	–	3–6		
			H16	0.2	6.0	–	125	–	2–4		
			H18	0.2	3.0	–	140	–	2–4		
1350	Al 99.50	Sheet, strip	O	–	–	–	–	–	25	A	BS 2897
			H4	–	–	–	95	–	8		
			H8	–	–	–	145	–	3		
3103	Al Mn1	Plate, sheet, strip	F	0.2	25.0	–	–	–	–	A	BS 1470
			O	0.2	6.0	–	90	–	20–25		
			H12	0.2	6.0	–	120	–	5–9		

Alloy	Designation	Form	Temper								Spec
			H14	0.2	12.5	—	140	—	3–7		
			H16	0.2	6.0	—	160	—	2–4		
			H18	0.2	3.0	—	175	—	2–4		
3105	Al Mn Mg	Sheet, strip	O	0.2	3.0	—	110	—	16–20	A	BS 1470
			H12	0.2	3.0	115	130	—	2–5		
			H14	0.2	3.0	145	160	—	2–4		
			H16	0.2	3.0	170	185	—	1–3		
			H18	0.2	3.0	190	215	—	1–2		
5083	Al Mg4.5 Mn	Plate, sheet, strip	F	3.0	25.0	—	—	—	—	A	BS 1470
			O	0.2	80.0	125	275	14	12–16		
			H22	0.2	6.0	235	310	—	5–8		
			H24	0.2	6.0	270	345	—	4–8		
5154A	Al Mg3.5	Sheet, strip	O	0.2	6.0	85	215	—	12–18	A	BS 1470
			H22	0.2	6.0	165	245	—	5–8		
			H24	0.2	6.0	225	275	—	4–6		
5251	Al Mg2	Plate, sheet, strip	F	3.0	25.0	—	—	—	—	A	BS 1470
			O	0.2	6.0	60	160	—	18–20		
			H22	0.2	6.0	130	200	—	4–8		
			H24	0.2	6.0	175	225	—	3–5		
			H28	0.2	3.0	215	255	—	2–4		
5454	Al Mg3 Mn	Plate, sheet, strip	F	3.0	25.0	—	—	—	—	A	BS 1470
			O	0.2	6.0	80	215	—	12–18		
			H22	0.2	3.0	180	250	—	4–8		
			H24	0.2	3.0	200	270	—	3–6		

(Continued)

Table 1.4 Continued

Alloy designation British Standard	Alloy designation ISO standards	Product forms	Temper designation	Thickness (mm) Over	Thickness (mm) Up to and including	Minimum 0.2% proof stress (N mm^{-2})	Minimum tensile strength (N mm^{-2})	Elongation (%) 5.65√S_0	Elongation (%) 50 mm	Durability rating	Related British Standard
6082	Al Si1 Mg Mn	Plate, sheet, strip	O	0.2	3.0	—	—	—	15–16	B	BS 1470
			T4	0.2	3.0	120	200	—	15		
				3.0	25.0	115	200	12	—		
			T6	0.2	3.0	255	295	—	8		
				3.0	25.0	240	295	—	8		
		Plate	T451	6.0	90.0	115	200	15	—		
		Plate	T561	6.0	90.0	240	295	8	—		
				90.0	115.0	230	285	7	—		
				115.0	150.0	220	275	6	—		
7020	Al Zn4.5 Mg1	Plate, sheet, strip	TB	0.2	25.0	170	280	10	12	C	BS 4300/14
			TF	0.2	25.0	270	320	8	10		

[a]Table compiled from BS 1470, BS 2897 and BS 4300/14.

Table 1.5 Mechanical properties of aluminium and aluminium extended bar, round tube and sections[a]

Alloy designation British Standard	Alloy designation ISO standards	Product forms	Temper designation	Thickness or diameter (mm) Over	Up to and including	Minimum 0.2% proof stress (N mm⁻²)	Minimum tensile strength (N mm⁻²)	Elongation (%) $5.65\sqrt{S_0}$	50 mm	Durability rating	Related British Standard
1050A	Al 99.50	All	–	–	–	–	60[b]	25[b]	23[b]	A	BS 1474
1200	Al 99.0	All	–	–	–	–	65[b]	20[b]	18[b]	A	BS 1474
1350	Al 99.50	All	–	–	–	–	60	25	23	A	BS 2898
			–	–	–	–	85	15	13		
5083	Al Mg4.5 Mn	All	–	–	150	125	275	14	13	A	BS 1474
			–	–	150	130[b]	280[b]	12[b]	11[b]		
5154A	Al Mg3.5	All	–	–	150	85	215	18	16	A	BS 1474
			–	–	150	100[b]	215[b]	16[b]	14[b]		
5251	Al Mg2	All	–	–	150	60[b]	170[b]	16[b]	14[b]	A	BS 1474
6061	Al Mg1 Si Cu	All	–	–	150	115	190	16	14	B	BS 1474
			–	–	150	240	280	8	7		
6063	Al Mg0.7 Si	All	–	–	200	–	140[c]	15	13	B	BS 1474
			–	–	200	–	100[b]	13[b]	12[b]		
			–	–	150	70	130	16	14		
			–	150	200	70	120	13	–		
			–	–	25	110	150	8	7		
			–	150	200	160	195	8	7		
6063A	Al Mg0.7 SiA	All	–	–	25	90	150	14	12	B	BS 1474
			–	–	25	160	200	8	7		

(Continued)

Table 1.5   *Continued*

Alloy designation British Standard	Alloy designation ISO standards	Product forms	Temper designation	Thickness or diameter (mm)		Minimum 0.2% proof stress (N mm⁻²)	Minimum tensile strength (N mm⁻²)	Elongation (%)		Durability rating	Related British Standard
				Over	Up to and including			$5.65\sqrt{S_0}$	50 mm		
6082	Al Si1 Mg Mn	All	–	–	25	190	230	8	7	B	BS 1474
				–	200	–	170[c]	16	14		
				–	200	–	110[b]	13[b]	12[b]		
				–	150	120	190	16	14		
				150	200	100	170	13	–		
				–	6	230	270	–	8		
				–	20	255	295	8	7		
				20	150	270	310	8	–		
				150	200	240	280	5	–		
7020	Al Zn4.5 Mg1	All	–	–	25	190	300	12	10	C	BS 4300/14
				–	25	280	340	10	8		

[a]Table compiled from BS 1474, BS 2898 and BS 4300/15. [b]Typical properties. [c]Maximum permitted value.

Table 1.6 Mechanical properties of aluminium and aluminium alloy drawn tube and seam-welded tube[a]

Alloy designation British Standard	Alloy designation ISO standards	Product forms	Temper designation	Wall thickness (mm) Over	Up to and including	Minimum 0.2% proof stress (N mm⁻²)	Minimum tensile strength (N mm⁻²)	Elongation (%) 5.65√S₀	50 mm	Durability rating	Related British Standard
1050A	Al 99.50	Drawn tube	O	—	12	—	95[b]	—	—	A	BS 1471
			H4	—	12	—	100	—	—		
			H8	—	12	—	135	—	—		
1200	Al 99.0	Drawn tube	O	—	12	—	105[b]	—	—	A	BS 1471
			H4	—	12	—	110	—	—		
			H8	—	12	—	140	—	—		
5083	Al Mg4.5 Mn	Drawn tube	O	—	10	125	275	12	—	A	BS 1471
			H2	—	10	235	310	5	—		
5154A	Al Mg3.5	Drawn tube	O	—	10	85	215	16	—	A	BS 1471
			H4	—	10	200	245	4	—		
5251	Al Mg2	Drawn tube	O	—	10	60	160	18	—	A	BS 1471
			H	—	10	175	225	5	—		
		Seam-welded tube	M	0.8	1.0	220	245	—	3	A	BS 4300/1
				1.2	2.0	220	245	—	5		
6061	Al Mg1 Si Cu	Drawn tube	TB	—	6	115	215	—	12	B	BS 1471
				6	10	115	215	—	14		
			TF	—	6	240	295	—	7		
				6	10	225	295	—	9		
6063	Al Mg0.7 Si	Drawn tube	O	—	10	—	155[b]	—	—	B	BS 1471
			TB	—	10	100	155	—	15		
			TF	—	10	180	200	—	8		

(Continued)

Table 1.6 *Continued*

Alloy designation British Standard	Alloy designation ISO standards	Product forms	Temper designation	Wall thickness (mm)		Minimum 0.2% proof stress (N mm^{-2})	Minimum tensile strength (N mm^{-2})	Elongation (%)		Durability rating	Related British Standard
				Over	Up to and including			5.65$\sqrt{S_0}$	50 mm		
6082	Al Si1 Mg Mn	Drawn tube	TB	–	6	115	215	–	12	B	BS 1471
				6	10	115	215	–	14		
			TF	–	6	255	310	–	7		
				6	10	240	310	–	9		

^{a}Table partly compiled from BS 1471 and BS 4300/1. ^{b}Maximum permitted value.

Table 1.7 Mechanical properties of aluminium and aluminium alloy ingots and castings[a]

Alloy designation British Standard	Alloy designation ISO standards	Product forms	Temper designation	Minimum tensile strength ($N\,mm^{-2}$)	Elongation[c] on $5.65\sqrt{}/S_0$ (%)	Durability rating	Related British Standard
LM2	Al Si10 Cu2	Gravity die cast	M	150	1–3	C	BS 1490
		Pressure die cast	M	150	–		
LM4	Al Si5 Cu3	Sand cast	M	140	2	C	BS 1490
		Gravity die cast	M	160	2		
		Sand cast	TF	230	–		
		Gravity die cast	TF	280	–		
LM5	Al Mg5 Si1	Sand cast	M	140	3	A	BS 1490
		Gravity die cast	M	170	5		
LM6	Al Si12	Sand cast	M	160	5	B	BS 1490
		Gravity die cast	M	190	7		
LM25	Al Si7 Mg0.5	Sand cast	M	130	2		
		Gravity die cast	M	160	3		
		Sand cast	TE	150	1		
		Gravity die cast	TE	190	2		
		Sand cast	TB7	160	2.5		
		Gravity die cast	TB7	230	5		
		Sand cast	TF	230	–		
		Gravity die cast	TF	280	2		

[a]Table partly compiled from BS 1490.

Table 1.8 Mechanical properties of aluminium and aluminium alloy forging stock and forgings[a]

Alloy designation British Standard	Alloy designation ISO standards	Product forms	Temper designation[b]	Thickness (mm) Over	Thickness (mm) Up to and including	Minimum 0.2% proof stress (N mm^{-2})	Minimum tensile strength (N mm^{-2})	Minimum elongation on 5.65$\sqrt{S_0}$ (%)	Durability rating	Related British Standard
1050A	Al 99.50	Forged	M	–	150	–	60	22	A	BS 1472
5083	Al Mg4.5 Mn	Forged, extruded	M	–	150	130	280	12	A	BS 1472
5154A	Al Mg3.5	Forged, extruded	M	–	150	100	215	16	A	BS 1472
5251	Al Mg2	Forged, extruded	M	–	150	60	170	16	A	BS 1472
6063	Al Mg0.7 Si	Forged, extruded	TB	–	150	85	140	16	B	BS 1472
			TB	150	200	85	125	13		
			TF	–	150	160	185	10		
			TF	150	200	130	150	6		
6082	Al Si1 Mg Mn	Forged	TB	–	150	120	185	16	B	BS 1472
		Extruded	TB	–	150	120	190	16		
			TB	150	200	100	170	13		
		Forged	TF	–	150	255	295	8		
		Extruded	TF	–	20	255	295	8		
			TF	20	150	270	310	8		
			TF	150	200	240	280	5		

[a]Table partly compiled from BS 1472. [b]Properties in M condition for information only.

have a well-defined yield point; however, typical stress–strain curves for some alloys are indicated in *Figure 1.6*.

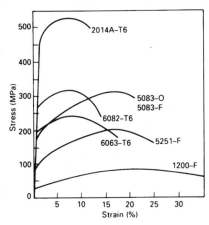

Figure 1.6 *Stress–strain curves for some aluminium alloys*

1.4 Physical properties

1.4.1 Density
The density of aluminium is approximately one-third that of steel and copper. The density of pure aluminium is 2.6898 g cm^{-3}. Trace amounts of alloying elements produce cast and wrought alloys in the density range of 2.66–2.80 g cm^{-3}.

1.4.2 Elastic modulus
Young's modulus is approximately one-third that of steel. For pure aluminium the value is 68.3 GPa. For high-strength copper bearing alloys the modulus may be as high as 74 GPa.

1.4.3 Thermal expansion
Pure aluminium has a coefficient of thermal expansion approximately twice that of steel at $23.5 \times 10^{-6}°C^{-1}$ in the range 10–100°C. Alloying elements have a small effect on this value. Figures in the range 20–$25 \times 10^{-6}°C^{-1}$ (for casting alloys) to 20–$24 \times 10^{-6}°C^{-1}$ for wrought alloys) are applicable.

1.4.4 Thermal conductivity
Aluminium is a good conductor of heat with a conductivity for pure aluminium of 244 W m^{-1}°C^{-1} (about 4.5 times that of steel). The thermal conductivity reduces with increased alloying to 109 W m^{-1}°C^{-1} for wrought alloys and to below 100 W m^{-1}°C^{-1} for some casting alloys.

1.4.5 Electrical resistivity and conductivity
Pure aluminium has a high electrical conductivity at 63% of IACS and an electrical resistivity of $2.69\mu\Omega$ cm at 20°C. These values increase with alloying to a resistance of over $6\mu\Omega$ cm at 20°C for some alloys and a conductivity down to below 28% IACS.

1.4.6 Melting point
Pure aluminium melts at 660°C but the melting point of some alloys may be as low as 530°C.

1.4.7 Non-magnetic properties
Aluminium and its alloys are virtually non-magnetic.

1.4.8 Reflectivity
For pure aluminium with a clean bright surface 80–85% of visible incident radiation and about 90% of heat will be reflected. Emissivity is typically 0.30 for a wavelength of 0.65μm.

1.4.9 Creep
Only commercially pure aluminium shows significant creep at room temperature. However, if the working temperature is held at around 200–250°C then the creep of some alloys will be significant.

1.4.10 Fatigue
Typical values of fatigue strengths at 50×10^6 cycles range from 20 M Pa for annealed commercially pure aluminium to 124 M Pa for 6082-TG material. Alloys with high manganese contents will exhibit higher values. Fatigue strength may be drastically reduced by unsatisfactory weld details.

1.4.11 Properties at elevated temperatures
Aluminium and its alloys show considerable reductions of strength at temperatures above 100°C. At 200°C the strength is approximately half that at room temperature; by 350°C most alloys will have lost most of their strength. Strength is not restored as temperature falls as the effect of work hardening will have been removed.

1.4.12 Properties at low temperatures
At temperatures below zero, aluminium and aluminium alloys exhibit higher tensile strengths and elongations than at room temperatures. No alloys suffer low-temperature brittleness and there is no point below which brittle fracture occurs.

1.5 Fire

Where fire resistance is required, sufficient insulation must be provided to prevent the temperature of the metal from rising above 200°C.

1.6 Durability

In normal exposure the quickly forming aluminium oxide film will provide a high resistance to corrosion. In coastal and marine environments an unprotected surface will roughen and acquire a grey stone-like appearance. In general, after an initial rapid weathering, little further change will be detected over periods of 20–30 years and perhaps longer.

A guide to the general protection of aluminium structures is shown in *Table 1.9*.

The metal may need additional protection of contacting surfaces in crevices. Additionally aluminium may be subject to electrochemical attack if in contact with other metals or the washings therefrom. *Table 1.10* gives guidance on the additional protection that may be required. The table indicates varying degrees of added protection on a scale 0 to 5. The scale is briefly explained below:

0 No additional treatment.
1 Heads of bolts/rivets may be painted with aluminium paint.
2 Contact surfaces cleaned and pretreated before assembly. Surfaces brought together while priming coat still wet. Heads of bolts/rivets and surrounding area painted. Generally no need to paint heads of stainless steel fasteners.
3 As 2 but additional protection by impervious tape insulant onto and extending beyond contact surfaces after priming coat has dried. Alternatives include use of neoprene washers or polysulphide elastomeric jointing compounds.
4 As 3 but heads of steel bolts, rivets and surrounding area around any junctions with steel or cast iron, not already treated, should be metal sprayed with aluminium. Chlorinated rubber or zinc-rich paint may be used but is not as durable as metal spray.
5 As 4 but with full electrical separation between adjacent but differing metals.

Contact with other non-metallic materials Aluminium will corrode if in contact with concrete, masonry or plaster and wood impregnated with certain preservatives. Soils, insulating materials and some chemicals may be harmful to aluminium. It is beyond the scope of this book to give full recommendations to avoid damage but each case should be individually examined to give the most appropriate protection.

1.7 Welding

Aluminium alloys may be welded. The processes used are fusion, resistance and solid-phase bonding. For general engineering purposes the normal process is fusion welding by the tungsten inert gas (TIG) or metal inert gas (MIG) process. *Table 1.11* gives guidance on the selection of weld filler metal for parent metal combinations.

Table 1.9 *General protection of aluminium structures*

Alloy durability rating	Material thickness (mm)	Requirement according to environment[a]								
		Atmospheric							Immersed	
		Rural	Industrial/urban		Non-industrial		Marine		Freshwater	Seawater
			Moderate	Severe	Moderate	Severe	Moderate	Severe		
A	All	None	None	P	None	P	None	P	None	None
B	<3	None	P	P	P	P	P	P	P	P
B	≥3	None	P[b]	P	None	P	None	P	P	P
C	All	None	P	P	P	P	P	P	P[c]	NR

[a]P, protection; NR, not recommended. [b]Requires only local protection to weld and heat-affected zones in urban non-industrial environments. [c]Not recommended if of welded construction.

Table 1.10 Additional protection at metal-to-metal contacts to combat crevice and galvanic effects

Metal joined to aluminium	Bolt or rivet metal	Procedure according to environment								
		Atmospheric				Marine			Immersed	
		Rural		Industrial/urban		Non-industrial	Industrial		Freshwater	Seawater
		Dry unpolluted	Mild	Moderate	Severe		Moderate	Severe		
Aluminium	Aluminium	0	0	0	2	2	0	2	0	2
	Steel, aluminized or galvanized steel, stainless steel	1	1	3	4	4	3	4	5	5
Zinc or zinc coated steel	Aluminium	0	0	2	2	2	2	2	2	5
	Steel, aluminized or galvanized steel, stainless steel	1	1	3	4	4	3	4	5	5
Steel, stainless steel, cast iron, lead	Aluminium	0	0	3	3	3	3	3	5	5
	Steel, aluminized or galvanized steel, stainless steel	1	1	4	4	4	4	4	5	5
Copper[a]	Aluminium	0	NR	NR	NR	NR	NR	NR	NR	NR
	Copper, copper alloy	0	3	5	5	5	5	5	5	5

[a]Contact surfaces and joints of aluminium to copper or copper alloys should be avoided if possible. If they are used, the aluminium should be of durability rating A or B, and the bolts and nuts should be of copper or copper alloy. NR, not recommended.

Table 1.11 Selection of weld filler metal for parent metal combinations

First part	Parent metal combination (second part)[a][b]							
	Al-Si castings LM2 LM4 LM6 LM25	Al-Mg castings LM5	3*** alloys 3103 3105	1*** alloys 1050A 1080A 1200 1350	7020	6*** alloys 6061 6063 6063A 6082	5*** alloys 5154A 5251 5454	5083
5083	NR[b]	Type 5	Type 5	Type 5	5556A	Type 5	Type 5	5556A
		Type 5	Type 5	Type 5	Type 5	Type 5	Type 5	Type 5
		Type 5	Type 5	Type 5	5556A	Type 5	Type 5	Type 5
5*** alloys								
5154A	NR[b]	Type 5	Type 5	Type 5	Type 5	Type 5	Type 5	
5251		Type 5	Type 5	Type 5	Type 5	Type 5	–[c]	
5454		Type 5	Type 5	Type 5	Type 5	Type 5	Type 5	
6*** alloys								
6061	Type 4	Type 4	Type 4	Type 4	Type 5	Type 4/5		
6063	Type 4	Type 5	Type 4	Type 4	Type 5	Type 4		
6063A	Type 4	Type 5	Type 4	Type 4	Type 5	Type 4		
6082								
7020	NR[b]	Type 5	Type 5	Type 5	5556A			
		Type 5	Type 5	Type 5	Type 5			
		Type 5	Type 5	Type 5	Type 5			
1*** alloys								
1050A	Type 4	Type 5	Type 4	Type 1[d]				
1080A	Type 4	Type 5	Type 3/4	Type 1[c]				
1200	Type 4	Type 5	Type 4	Type 1[d]				
1350								
3*** alloys	Type 4	Type 5	Type 3[d]					
3103	Type 4	Type 5	Type 3					
3105	Type 4	Type 5	Type 3[d]					
Al-Mg castings	NR[b]	Type 5						
LM5		Type 5						
		Type 5						
Al-Si castings	Type 4							
LM2 LM4	Type 4							
LM6 LM25	Type 4							

	Filler metal groups		Durability
Type 1	1080A	1050A	A
Type 3	3103		A
Type 4	4043A	4047A	B
Type 5	5056A	5356 5556A 5183	A

[a]Filler metals for parent combination to be welded are shown in one group, which is located at the intersection of the relevant parent metal row and column. In each group, the filler metal for maximum strength is shown in the top line: in the case of 6*** and 7020 alloys, this will be below the fully heat-treated parent metal strength. The filler metal for maximum resistance to corrosion is shown in the middle line. The filler metal for freedom from persistent weld cracking is shown in the bottom line. [b]NR, not recommended. The welding of alloys containing approximately 2% or more of MG with Al-Si (5–12% Si) filler metal (and vice versa) is not recommended because sufficient Mg$_2$Si precipitate is formed at the fusion boundary to embrittle the joint. [c]The corrosion behaviour of weld metal is likely to be better if its alloy content is close to that of the parent metal and not markedly higher. Thus for service in potentially corrosive environments it is preferable to weld 5154A with 5154A filler metal or 5454 with 5554 filler metal. However, in some cases this may only be possible at the expense of weld soundness, so that a compromise will be necessary. [d]If higher strength and/or better crack resistance is essential, type-4 filler metal can be used. [e]For welding 1080A to itself, 1080A filler metal should be used.

Table 1.12 *The general availability of product forms[a]*

Alloy	Plate	Sheet and strip	Extruded sections				Drawn tube	Longitudinally welded tube	Forgings	Castings
			Solid bar and simple sections	Complex sections	Round and rectangular tube	Hollow section				
1050A	A	A	A	–	–	–	A	–	A	–
1080A	A	A	A	–	–	–	A	–	–	–
1200	MS	MS	A	–	–	–	A	–	–	–
1350	–	A	A	–	A	–	–	–	–	–
3103	MS	MS	–	–	–	–	–	–	–	–
3105	–	M	–	–	–	–	–	–	–	–
5083	MS	MS	A	–	–	–	A	–	M	–
5154A	M	M	A	–	–	–	A	–	M	–
5251	M	MS	A	–	–	–	A	–	M	–
5454	M	M	A	–	–	–	–	MS	–	–
6061	A	A	M	M	M	M	M	–	M	–
6063	–	–	MSX	MS	MS	M	M	–	M	–
6063A	–	–	M	M	M	M	–	–	–	–
6082	MS	MS	MSX	M	MS	M	M	–	M	–
7020	A	A	A	–	–	–	–	–	–	–
LM2	–	–	–	–	–	–	–	–	–	M
LM4	–	–	–	–	–	–	–	–	–	M
LM5	–	–	–	–	–	–	–	–	–	M
LM6	–	–	–	–	–	–	–	–	–	M
LM25	–	–	–	–	–	–	–	–	–	M

[a]MS, standard product manufactured to order with a limited range available from stock; M, standard product manufactured to order; A, special product manufactured by special arrangement; X, includes sections from BS 1161.

1.8 Standard products and availability of materials

The range of aluminium and aluminium alloys given in *Table 1.3* is not produced in all wrought and cast forms. The general availablity of product forms for each alloy is indicated in *Table 1.12* and the range of sizes of extruded sections to BS1161 is contained in *Table 1.13*.

An indication of the ISO standards most appropriate to aluminium and aluminium products is given in *Table 1.14*.

Bibliography

BAYLEY, M.J. and BUSLON, P.S. In DORAN, D.K. (ed.), *Construction Materials Reference Book*, Butterworth-Heinemann, Oxford, Chapter 2 (1992)

Table 1.13 *BS 1161 standard extruded sections—range of sizes*

Section type	Size range (mm)
Equal angles	30 × 30–120 × 120
Unequal angles	50 × 38–140 × 105
Channels	60 × 30–240 × 100
T-sections	50 × 38–120 × 90
I-sections	60 × 30–160 × 80
Equal bulb angles	50 × 50–120 × 120
Unequal bulb angles	50 × 37.5–140 × 105
Lipped channels	80 × 40–140 × 70
Bulb T-sections	90 × 75–180 × 150

Table 1.14 ISO standards for aluminium and aluminium alloy materials and products

Type and content of standard	Castings	Forgings	Wire rod	Drawn wire	Drawn products: rods, bars and tubes	Extruded products: rods, bars, tubes, profiles	Rolled products: sheet, strip and plates
Terminology	ISO 3134-1 ISO 3134-4	ISO 3134-1 ISO 3134-3	ISO 3134-1 ISO 3134-3	ISO 3134-1 ISO 3134-3	ISO 3134-1 ISO 3134-3	ISO 3134-1 ISO 3134-3	ISO 3134-1 ISO 3134-3
Designation of alloys	ISO 2092	ISO 2092	ISO 2092	ISO 2092	ISO 2092	ISO 2092	ISO 2092
Form of product Composition	–	– –	– –	ISO 209-2 ISO 209-1	ISO 209-2 ISO 209-1	ISO 209-2 ISO 209-1	ISO 209-2 ISO 209-1
Temper designation	ISO 2107	ISO 2107	ISO 2107	ISO 2107	ISO 2107	ISO 2107	ISO 2107
Technical delivery conditions	ISO 7722	–	–	ISO 6365-1	ISO 6363-1	ISO 6362-1	ISO 6361-1
Mechanical and physical properties	ISO 3522	–	–	–	–	ISO 6362-2	ISO 6361-2
Dimensions and tolerances	–	–	–	–	Round bars ISO 5193 and ISO 7274	Profiles ISO 6362-4	Strips ISO 6361-3 Sheets and plates ISO 6361-4

–, No ISO standard.

2 Cast iron

The most fundamental difference between cast iron and either wrought iron or steel lies in the higher carbon content of the former. The broad range of values for this proportion are given in *Table 2.1*.

Table 2.1 *Carbon contents of cast iron, steel and wrought iron*

Material	Carbon content (%)
Cast iron	2.0–4.5 (generally 2.5–4.0)
Steel	0.2–1.0
Wrought iron	0.02–0.05

Cast iron used in the construction industry may be divided into three main categories:

(1) 'Historic' cast iron, that is mainly grey cast iron, as widely used in structures between about 1780 and 1880 but also including some malleable cast iron, made by the heat treatment of white iron castings.
(2) Modern grey cast iron, which is virtually the same as the historic grey iron but is generally of a higher quality and is covered by British Standards. It is mainly used in mechanical engineering rather than in structures.
(3) Ductile cast iron, or spheroidal graphite cast iron, which again is little used in construction today but which is covered by British Standards and could have a major future as a structural material. It is a relatively modern material dating from after 1946.

In discussing the properties and uses of cast iron it is convenient to adhere to these categories, although there must be some overlapping between the three.

Category (1), historic iron, is only really of interest to civil and structural engineers in connection with the appraisal, repair and adaptation of existing structures. The most notable characteristics of this type of iron are its much greater strength in compression than in tension and its non-linear behaviour under tensile load. These features present problems of analysis which are not shared by the other metals normally used for structures.

Although the main outlet for category (2), modern grey cast iron, is in mechanical equipment it is still used extensively on the municipal side of civil engineering for pipes, pipe fittings, bollards, manhole covers, gratings and other applications where simple robustness is needed rather than a calculable structural performance. Nevertheless, in all these fields it is liable to be superseded to a large extent by category (3), ductile cast iron.

The most important features of ductile cast iron are that it is comparably strong in tension and in compression and that its strength and stiffness are similar to those of rolled steel, yet it can be moulded into complex shapes with the same freedom as grey cast iron.

As with other materials formed by similar methods, cast iron may sometimes be identified by surface evidence of the casting process.

The basic properties of these three categories of cast iron are shown in *Tables 2.2, 2.3, 2.4, 2.5* and *2.6*. It should also be noted that carbon content has a marked effect on the melting point of cast iron as illustrated in *Figure 2.1*. Typical stress–strain relationships are shown in *Figure 2.2*.

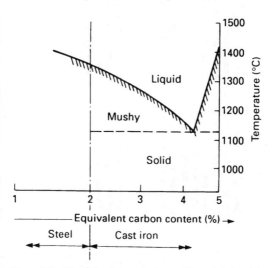

Figure 2.1 *Effect of carbon content on melting point of cast iron*

Table 2.2 Type of historic (mainly 19th century) cast iron

Type of cast iron	Microstructure	Physical properties	Uses	Notes
Grey	Graphite in flake form in an iron matrix. Flakes form discontinuites	Strong in compression. Relatively weak in tension. Good resistance to corrosion. Easily machined and cut. Very large castings practicable. Limited ductility	Main form of cast iron used in construction, for columns, beams, decorative panels, etc., as well as machinery	Historic cast iron nearly all grey iron. Little used in construction today except for pipes, pipe fittings, manhole covers, etc.
White	No free graphite. Carbon combined with iron as hard carbides. Low equivalent carbon. Low silicon content	Very hard and very brittle. Machined by grinding only	Surfaces needing high resistance to abrasion	Virtually irrelevant to construction industry
Malleable	Made by prolonged heat treatment of white iron castings. Carbides transformed into graphite in nodular form with few discontinuities in iron matrix	Very strong in tension as well as compression, with good ductility	Hinges, catches, step irons and similar castings of limited size. Decorative panels of fragile design	Likely to be superseded by ductile iron which has similar properties and can be cast in a wide range of section thicknesses

Table 2.3 Results of flexural tests on different sizes of grey cast iron beam and on small specimens cut from beams tested

Test set (Ref.)	Date of test	Shape of beams	Depth of section (mm)	Results of flexural tests on beams or bars				Results of flexural tests on 25 mm × 25 mm specimens cut from beams tested				X/Y
				No. of specimens	Modulus of rupture (N mm⁻²)			No. of specimens	Modulus of rupture (N mm⁻²)			
					High	Low	Mean X		High	Low	Mean Y	
A* (12 + 13)	Various, 19th century mainly	Square	25	Many	400	280	340	–	–	–	–	–
		Rect. 25 wide	75	1 (?)	–	–	approx. 258	–	–	–	–	–
		Rect. 50 wide	75	1 (?)			250					
		square	75	1 (?)			200					
B (12 + 13)	Various 19th century	Small I or inverted T	100 → 175	7	203	143	175	–	–	–	–	–
C (14)	1984	Inverted T	333	2	210	156	183	–	–	–	–	–
D* (15)	1834	I	298	6	190	138	156	12	464	394	427	2.7
E* (16)	1987	I	360	1	–	–	122	5	250	218	237	1.9

(Continued)

Table 2.3 Continued

Test set (Ref.)	Date of test	Shape of beams	Depth of section (mm)	Results of flexural tests on beams or bars					Results of flexural tests on 25 mm × 25 mm specimens from beams tested					X/Y
				No. of specimens	Modulus of rupture (N mm^{-2})				No. of specimens	Modulus of rupture (N mm^{-2})				
					High	Low	Mean			High	Low	Mean		
							X					Y		
F* (17)	1944	1	298 → 634	12	135	116	122	9		390	252	332		2.7
G1† (18)	1968	I	460 & 675	2 } 4	118	101	108	–		–	–	–		–
G2† (18)	1968	I	460	2	159	158	158.5	–		–	–	–		–

*D–F: 358 N mm^{-2} mean for wt 25 mm × 25 mm specimens, close to results for set A.
†Beams in set G2 noted as superior to those in set G1.

Table 2.4 *Limits of permissible stresses for cast iron in BD21/84*

	All permanent loads or most favourable combinations of permanent and live load	All live load
Compression	154 N mm⁻² (10 ton in.⁻²)	80 N mm⁻² (5.2 ton in.⁻²)
Tension	46 N mm⁻² (3 ton in.⁻²)	24.5 N mm⁻² (1.6 ton in.⁻²)
Shear	46 N mm⁻² (3 ton m⁻²)	See explanation in BD21/84

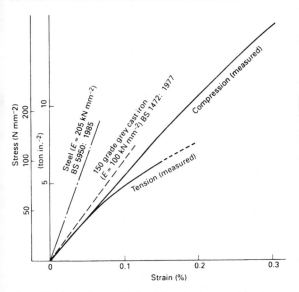

Figure 2.2 *Typical stress–strain relationship of grey cast iron in tension and compression. (From Clarke, 1850)*

References

CLARKE, E., *The Britannia and Conway Tubular Bridges*, Day and Son, London (1850)

UNWIN, W.C., *The Testing of Materials of Construction*, Longman, London (1910)

Bibliography

SUTHERLAND, R.J.M. In DORAN, D.K. (ed.), *Construction Materials Reference Book*, Butterworth-Heinemann, Oxford, Chapter 3 (1992)

Table 2.5 Main structural properties of grey cast iron (partly from BS 1452: 1977)

Basis of acceptance		For guidance only (quoted in BS 1452: 1977 based on BCIRA data)				
Grade No.	Tensile strength ($N\,mm^{-2}$)	0.1% Proof stress tension ($N\,mm^{-2}$)	Compressive strength ($N\,mm^{-2}$)	0.1% Proof stress compression ($N\,mm^{-2}$)	Shear strength ($N\,mm^{-2}$)	Elastic modulus ($N\,mm^{-2}$)
150	150	98	600	195	173	100×10^3
180	180	117	672	234	207	109×10^3
220	220	143	768	286	253	120×10^3
260	260	169	864	338	299	128×10^3
300	300	195	960	390	345	135×10^3
350	350	228	1080	455	403	140×10^3
400	400	260	1200	520	460	145×10^3

Table 2.6 Tensile strength of grey cast iron (from Unwin, 1910)

Experimenter	No. of tests	Cross-section (in.²)	Tenacity (ton in.⁻²)			Probable No. of fusion	Condition of test bars	Reference
			H1ghest	Lowest	Mean			
Minard and Desormes	13	0.23–0.5	9.08	5.09	7.19	–	Rough	Love, *Résistance de la Fonte* (1815)
Hodgkinson and Fairbairn	–	1–4	9.76	6.00	7.37	2nd	Rough	Brit. Assoc. Rep. VI, (1837)
	81	3–4.5	10.5	4.9	6.83	2nd	Rough	Report on App. Iron (1849)
Woolwich	53	–	15.3	4.2	10.4	2nd and 3rd	?	Report of 1858 (1856)
	6	–	–	–	13.7§	–	–	Report on Metal for Cannon (1856)
Wade	4	–	–	–	9.1†	–	–	Report on Metal for Cannon (1856)
Turner‡	–	1.0	15.7	4.75	–	2nd	Rough	J. Chem. Soc. (1885)
Rosebank foundry	23	–	18.2	6.5	15.3*	–	–	Industries (April 1887)
Unwin	6	0.75	–	–	13.7	2nd	Turned	–
	3	1.0	17.3	14.9	15.7	2nd	Turned	–
Wade	–	–	20.5‖	–	–	–	–	–

* Selected as good iron.
† Selected as bad iron.
‡ Special series of experimental test bars, with varying proportion of silicon.
§ Mean of ten best specimens.
‖ Highest result obtained.

3 *Wrought iron*

3.1 Introduction

Compared with cast iron, wrought iron is a simple material. Its structural properties are effectively the same in tension, compression and bending and far more certain than those of cast iron. It is a ductile material free from the 'brittleness' of grey cast iron and compared with historic grey cast iron about two to three times as strong in tension.

Between the mid-1840s and the early 1850s wrought iron gradually took over from cast iron as the high-performance structural material and was recognized as such until towards the end of the 19th century when, in turn, it was superseded by mild steel, a material of greater all-round strength capable of being formed industrially into much larger structural sections.

Today there is no wrought iron industry at all in Britain, although attempts are being made to reintroduce it on a museum level. Even though in most respects today's steel is a superior material there are some properties of wrought iron which are better than those of steel, notably its greater resistance to corrosion and the ease with which it can be shaped or 'wrought' into complex shapes of varying section.

Most of the references to tests and many of the opinions quoted date from the 19th or early 20th centuries. The reason for this is that virtually no research has been carried out on wrought iron in the last 75–100 years and very little has been written on the subject except of the most general nature. Appraising engineers should bear this time-scale in mind, especially when making comparisons with more recent experience with steel. When in doubt they may feel the need to carry out their own confirmatory tests.

3.2 Types of wrought iron

There are two basic types of wrought iron.

(1) *Blacksmith's wrought iron.* Before about 1500 AD and the introduction of the blast furnace, virtually all iron was of this type. It was extracted in a pasty, never molten, state and hammered into required shapes. This eventually gave way to wrought iron made from cast iron which had been produced in a finery or a puddling furnace.

(2) *Industrialized wrought iron.* This type became available following the introduction of Cort's patent of 1783 for the puddling furnace.

3.3 Properties

Wrought iron is almost pure iron having a carbon content of only 0.02% to 0.05%. It is formed mainly by hammering or rolling. It has a fibrous texture not dissimilar to timber which is most notable at fracture points or when the material is badly corroded. When polished and etched the presence of slag lines is detectable. There is no British Standard for the material, which presents problems for those appraising structures where strength is critical.

In the absence of a standard the most useful official guidance is to be found in the Department of Transport's document BD 21/84 'The assessment of highway bridges'. Historical research has revealed a range of test results dating back to 1837 and these are summarized in *Tables 3.1, 3.2* and *3.3*.

Figures quoted for elastic modulus are in the range of 170 kN mm^{-2} to 220 kN mm^{-2} with a mean of approximately 195 kN mm^{-2}. Typical stress/strain curves derived from an early book by W. C. Unwin are shown in *Figure 3.1*. As is typical with metals the effect of cold working of wrought iron is significant. Two examples of this phenomenon are shown in *Table 3.4*. Shear strengths for wrought iron are not well documented, although E. Clarke has quoted figures of 370 N mm^{-2} (single shear), 340 N mm^{-2} (double shear) from tests carried out around 1840. These values are for rivet iron with a tensile strength of 340 N mm^{-2}. Others have quoted shear strengths

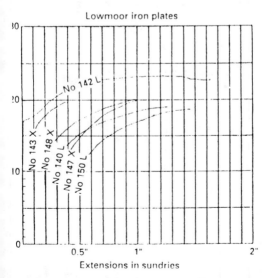

Figure 3.1 *Typical stress–strain curves for wrought iron plate. (From Unwin, 1910)*

Table 3.1 Recorded results of tests on strength and ductility of wrought iron bars, flats, angles, etc. (see Table 3.3 for wrought iron plate)

Author and date of publication	No. of tests	Ultimate tensile strength		Yield stress		Elongation at failure (%)	Notes
		Range (N mm⁻²)	Mean (N mm⁻²)	Range (N mm⁻²)	Mean (N mm⁻²)		
Barlow, 1837	9 (yield 6)	420–491	448	221–329	272	0.5–12.1 (mean 5.1)	Telford/Brunton test
	7	407–482	453	–	–		S. Brown tests
	10	387–526	464	278–417	340	–	M. Brunel tests (best iron)
	10	464–557	495	340–433	371	–	M. Brunel tests (best best iron)
	6	417–510	479	–	–	–	M. Brunel (best iron)
	3	155–186	–	–	170	–	P. Barlow (not to destruction)
Hodgkinson, 1849	1	368	368	195	195	3.5	Fixing failed early — Taken from plots of stress/strain curves
	1	–	–	200	200	–	
Humber, 1870	6	285–386	346	–	–	–	Ship straps, angles, etc.
	4	283–303	291	166–185	171	–	Very soft Swedish iron
	188	308–476	397	–	–	–	Bars ⎫ Kirkaldy tests
	72	261–440	378	–	–	–	Angle irons ⎭
Unwin, 1910	12	348–475	387	170–216	196	–	Steel Committee tests
Cullimore, 1967	10	278–352	322	195–258	223	–	Flat tie bars from Chepstow railway bridge
Sandberg, 1986	4	327–340	336	227–252	239	–	All from the same flat tie member
Overall mean			396		260		

Table 3.2 *Typical recorded results of tests on strength and ductility of wrought iron plate compared with test results for bars, angles, etc.*

Author and date of publication	No. of tests	Direction of loading in relation to grain	Ultimate tensile strength		Yield stress		Elongation at failure	Notes
			Range ($N\,mm^{-2}$)	Mean ($N\,mm^{-2}$)	Range ($N\,mm^{-2}$)	Mean ($N\,mm^{-2}$)	(%)	
Fairbairn, 1850	11	Along	312–407	348	–	–	–	Plate 6–7 mm thick
	8	Across	286–425	356	–	–	–	Strength very similar in both directions
Clark, 1850	12	Along	278–340	303	–	–	0.6–12.5	Plate 12–17 mm thick
	2	Along	304–312	308	–	–	–	Strengths on average
	2	Across	258–262	260	–	–	–	14% lower across grain than along it
Humber, 1870	16	Along	299–404	355	–	–	–	Strength on average 8.2%
	16	Across	285–380	326	–	–	–	Lower across gain than along it
Unwin, 1910	51	Along	336–443	377	185–281	240	6.4–30.9	Tests from Bohme, Berlin, 1884
	13	Across	307–360	325	200–237	215	1.7–16.9	UTS* averages 14% less across grain and yield 10% less than along it
Warren, 1894	1	Along	338	338	165	165		
	1	Along	324	324				
	1	Across	280	280				
Overall mean along grain				358	⎱ Taken together strength across gain as disclosed by these tests			
% Reduction on strength of bars, etc. *(Table 3.1)*				9.6	⎰ averages 8.7% less than that along the grain			
Overall mean across grain				327				
% Reduction on strength of bars, etc. *(Table 3.1)*				17.4				

*UTS, ultimate tensile strength.

Table 3.3 Typical figures for the strength and ductility of wrought iron as published (or required) at different dates but not necessarily related directly to any tests (see also Tables 3.1 and 3.2)

Author and date	Ultimate tensile strength ($N\,mm^{-2}$)	Yield stress ($N\,mm^{-2}$)	Elongation at failure (%)	Notes
Clark, 1850	Bars, etc. 371 Plate along grain 309	185	–	Records 'useful' strength in tension and compression as 185 $N\,mm^{-2}$ and cites bar sinking in compression at 232 $N\,mm^{-2}$
Matheson, 1873	Bars, etc. 309–371 Plate 278–340	'Permanent set' likely at 155–170	–	
Humber, 1870	Bars 387 (325 compression) Plates 340 (278 compression)	–	–	
Warren, 1894	Bars, angles, etc. 309–371 (varies slightly with size)	185	–	
British Standard SI-1939 (withdrawn)	Bars, angles, etc. 309–387 Plates with grain 309–371 Plates across grain 263–375	–	14–33 depending on grade 6–14 depending on grade and thickness 3–5 depending on grade and thickness	
BD21/84: D.o.T.* basis for assessment of bridges, 1984	–	220 characteristic yield strength given for use in assessment		

*D.o.T., Department of Transport.

Table 3.4 Effect of cold working on the properties of wrought iron

Form of iron	Type of cold working	Ultimate tensile strength ($N\,mm^{-2}$)	Elastic limit ($N\,mm^{-2}$)	Elongation to failure (%)	Source
Bar	Hot rolled to 32 mm diameter	402	–	20.3	Unwin, 1910, quoting from Fairbairn
	Subsequently cold rolled down to 25 mm diameter	594	–	8.0	
Plate	Hot rolled to 8 mm thick	367	224	15.0	Unwin, 1910, quoting from Considere
	Subsequently cold rolled down to 7.1 mm thickness	460	408	7.0	
Wire	Hot rolled	449 max. 283 min. 366 mean (approx.)	–	–	Rankine, 1872
	Subsequently	787 max. 490 min. 635 mean (approx.)	–	–	

in the order of 67% to 77% of tensile strengths for different irons. The coefficient of linear expansion is similar to cast iron and steel and is in the range 10×10^{-6} and 12×10^{-6} per °C at normal working temperatures. As with cast iron and steel the strength of wrought iron drops sharply above 400°C. The durability of wrought iron is generally accepted to be greater than steel but less than cast iron.

3.4   Repairs to wrought iron

Historically, wrought iron was jointed either by forge welding or riveting. As both these techniques have been largely abandoned it is necessary to resort to bolting or electric-arc welding. The former, used for repair, in conjunction may be unsightly, the latter does require a high degree of skill. It is therefore essential to seek expert advice in executing such repairs; furthermore it may be necessary to carry out extensive trials before arriving at an optimum solution.

References

BARLOW, P., *A Treatise on the Strength of Timber, Cast Iron, Malleable Iron and Other Materials*, John Weale, London (1837)

CLARK, E., *The Britannia and Conway Tubular Bridges*, Day and Son, London (1850)

CULLIMORE, M.S.G., Fatigue strength of wrought iron after weathering in service, *The Structural Engineer*, **45** (5) (May 1967)

FAIRBAIRN, W., *Iron: Its History, Properties and Processes of Manufacture*, A. & C. Black, Edinburgh (1869)

HODGKINSON, Evidence to the Royal Commission on the application of iron to railway structures (1849)

HUMBER, W., *Cast and Wrought Iron Bridge Construction*, Lockwood, London (1870)

MATHESON, E., *Works in Iron, Bridge and Roof Structures*, E. & F.N. Spon, London (1873)

RANKINE, W.J.M., *A Manual of Civil Engineering*, C. Griffin, London (1872)

SANDBERG, Private communication (1986)

UNWIN, W.C., *The Testing of Materials of Construction*, Longman, London (1910)

WARREN, W.H., *Engineering Construction in Iron, Steel and Timber*, Longman, London (1894)

Bibliography

SUTHERLAND, R.J.M.   In   DORAN, D.K.   (ed.), *Construction Materials Reference Book*, Butterworth-Heinemann, Oxford, Chapter 4 (1992)

4 Steel

4.1 Introduction

The term 'steels' encompasses a multitude of materials. Smithell's *Metals Reference Book* contains some 62 pages devoted to steels—approximately 600 combinations of composition and treatment are listed.

Constructional steels are a subset whose limits are defined by:

(1) Availability in large quantity at acceptable cost
(2) Ability to be fashioned into suitable sectional shapes
(3) Acceptable strength
(4) Acceptable toughness
(5) Ability to be welded without the need for stringent precautions.

The main group of materials that meet these requirements are alloys of iron with carbon and various other alloying elements, principally manganese. Manganese is always present in amounts between 0.5 and 1.5%. Weldability requirements place an upper limit on the carbon content at about 0.25%; high-strength fasteners may contain more. Other alloying elements may be introduced deliberately or be irremovable impurities. Each may act alone in determining the properties of the steel. More frequently, the metallurgy is dominated by interactions between the constituents of the steel.

The quality control of steels produced in the first half of this century was by control of composition and avoidance of obvious manufacturing defects. The introduction of all welded structures in the 1940s gave rise to a series of problems caused by fast brittle fractures. These were overcome by tighter control of composition and of metallurgical features such as grain size. More recent developments of high-strength low-alloy (HSLA), or microalloyed, steels demand much more stringent controls over both composition and structure. Tightly controlled rolling schedules are necessary for this purpose. Current research is aimed at producing the same quality of steel with less demanding control over rolling.

The overall production route for steels is summarized in *Figure 4.1*. Most steel now made in the UK is derived from two raw materials: liquid pig iron and scrap. In 1984, 15.12 million tonne of steel was made in the UK, using 7.86 million tonne of scrap.

The refining methods which convert these materials into steel have changed considerably since Bessemer's invention (announced in 1856) and the development of the open-hearth

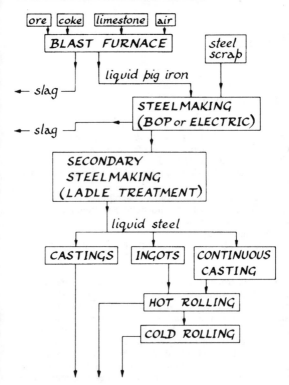

Figure 4.1 *Scheme to show the routes of production for steels (BOP, basic oxygen process)*

method (introduced in 1865). Since about 1970, most Bessemer and open-hearth furnaces have been discarded in favour of the basic oxygen process and related processes and the electric arc furnace. Much steel made by the older methods is still in use.

In addition to 'structural' steels, this chapter also includes steel for the reinforcement of concrete. In summary these are:

(1) Carbon steel bars for the reinforcement of concrete (BS 4449: 1988).
(2) Cold reduced wire (BS 4482: 1985) and welded steel fabric for the reinforcement of concrete (BS 4483: 1985).
(3) Pre-stressing steel wire and strand (BS 5896: 1980) and bars (BS 4486: 1988) for the reinforcement of concrete.

4.1.1 Coding system for steel
BS 4360: 1990 lays down a coding system. The code consists of a number representing the ultimate tensile strength of the steel in hectobars (1 hectobar = 10 MPa = $10 \, \text{N} \, \text{mm}^{-2}$). This is

followed by a letter representing the impact performance of the steel (see below).

Grade	Ultimate tensile strength (MPa)	
	1979 edition	1986 and 1990 editions
40	400–480	340–500
43	430–540	430–580
50	490–620	490–640
55	550–700	550–700

4.2 Mechanical forming processes

Steel may be formed into usable profiles by one of the following processes:

(1) Casting
(2) Rolling
(3) Forging
(4) Seamless tube forming
(5) Drawing.

These are dealt with, briefly, below.

4.2.1 Casting
Most steel compositions can be cast directly into complex shaped components. Examples include stern frames for ships and nodal regions for large structures (oil rigs, etc.).

Solidification shrinkage is large, but good process design can avoid porosity and other defects in the finished casting. The castings are usually heat treated to relieve any residual stresses caused by uneven cooling and to refine the grain structure of the material.

In general, acceptable yield strengths and tensile strengths are obtained, although toughness is often poor when compared with wrought counterparts.

British Standards for general engineering purposes are listed in BS 3100: 1991.

4.2.2 Rolling
Rolling is the most widely used process; it can be done hot or cold. A wide range of cross-sectional shapes can be produced. *Figure 4.2* clarifies some of the jargon used to describe semi-finished and finished products.

Hot rolling All constructional steels are hot rolled to begin with; for many products, especially those with large metal thicknesses, this is the only stage. Ingots or continuously cast slabs are preheated to 1200–1300°C to make them soft enough to deform and shape. The temperature falls as processing is

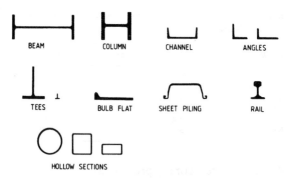

Figure 4.2   *The principal structural sections*

continued and preheating to lower temperatures is often necessary at intermediate stages. The total amount of work done by
rolling and the finishing temperatures are important in the
metallurgical control of quality. For example, BS 11; 1985 for
railway rails requires at least 8:1 reduction in cross-sectional
linear dimensions by hot rolling. In general low finishing temperatures (*c.* 900°C) are required. To give very fine grain sizes,
which means a good combination of strength and toughness,
controlled rolling may be used. This entails close control over
finishing temperatures, which are usually close to 850°C, but
may be as low as 650–750°C.

Cold rolling   Cold rolling cannot be used to the same degree as
hot working for shaping purposes, because of the increased
strength of the cold workpiece. Modest reductions can be
achieved by rolling to give better size tolerances and surface
qualities. It is used mainly for lightweight sections. Work hardening during the process gives increases in yield strength at the
expense of ductility and toughness. Profiling by bending, folding, etc., into corrugated or other shapes may be used to add
section stiffness to thin strip or sheet products (*Figure 4.3*).

4.2.3 Forging
Hot forging is used to make shaped, usually blocky, artifacts
which need to have properties better than those of castings.
Presses or various forms of stamps, hammers, etc., are used
to shape hot steel and to modify coarse as-cast structures into
more homogeneous and finer grained wrought structures.

4.2.4 Seamless tube forming
Seamless tubing is produced by hot deformation of initially
cylindrical billets. Rolls which generate a spinning motion
about the billet axis as well as motion along the billet axis are
used to force the billet around a mandrel which defines the
internal diameter of the tube.

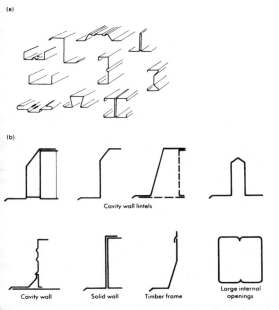

(a)

(b)

Cavity wall lintels

Cavity wall   Solid wall   Timber frame   Large internal openings

Figure 4.3 *(a) Typical cold-formed section profiles. (Reproduced by kind permission of the Director, The Steel Construction Institute.) (b) Sections of some of the steel lintels used in domestic housing. (Reproduced from* Metals and Materials, *the journal of the Institute of Metals, Vol. 4, No. 6, 1988, p. 357)*

4.2.5 Drawing
In wire drawing, a feedstock is pulled through a die by the product. If the material did not work harden, the process would not work. Successive reductions through many dies can be used to impart very high strengths by work hardening. The price paid is reduction in ductility and toughness.

In addition to the basic processing indicated above other hollow sections are made by folding and welding along the length of the part. Some tube is made by coiling strip in a helical fashion and welding along the helical seam.

4.3 Available sections

The range of sections available for use in construction is considerable. The Steel Construction Institute has produced a most comprehensive guide to structural sections: '*Steelwork Design: Guide to BS 5950: Part 1: 1985. Volume 1: Section Properties Member Capacities*' (2nd Ed. 1987). This deals with the following range of sections:

- Universal beams and columns
- Joists
- Bearing piles
- Hollow sections (circular, square, rectangular)
- Channels
- Angles (equal and unequal)
- Castellated members
- Tees.

Member capacities are indicated for grade 43 and grade 50 steel.

Further information on cold-formed sections is available from manufacturers.

Steel for the reinforcement of concrete is available in bar form for normal reinforced concrete and either in wire/strand or bar for prestressed concrete.

For normal reinforced concrete the basic types of bar available in the UK and complying with BS 4449 are:

(1) Grade 250 bars in plain round form (smaller sizes only)
(2) Grade 460 bars in a ribbed profile complying with the bond classification type 2 (defined in BS 4449 by projected rib area or by a pull-out test).

Stainless-steel bars are also available to BS 6744: 1986 standard—'Austenitic stainless steel bars for the reinforcement of concrete'—with properties similar to bars suitable to BS 4449: 1988. A British Standard is also in preparation covering fusion bonded epoxy coated reinforcement.

4.3.1 Sizes and lengths (reinforcing bars)

It should be noted that the word 'size' is used in preference to 'diameter'. 'Size' implies 'nominal size', that is the diameter of a circle with an area equal to the effective cross-sectional area of the bar. With a ribbed bar it is not possible to measure a cross-dimension which equals the size of the bar because the circle circumscribing the ribs will have a diameter greater than the nominal size.

The cross-sectional area derived from the mass per metre run using an assumed density of $0.00785 \, \text{kg mm}^{-2}$ per metre run is given in *Table 4.1*. Grade 250 plain round bars are available in sizes 8, 10, 12 and 16 mm. Grade 460 bars are commonly available and stocked in sizes 8, 10, 12, 16, 20, 25, 32 and 40 mm. Sizes 6 and 50 mm may be obtained by special arrangement, but they are rolled to order and their use can result in delay.

The standard length of bars cut at the steel mill and available from suppliers is 12 m. However, sizes 6, 8 and 10 mm are sometimes stocked in 6, 9 or 10 m lengths. The maximum length of bar obtainable by special arrangement and with adequate notice is 18 m. Longer lengths can be produced by the use of bar couplers.

Table 4.1 *The cross-sectional area*[a] *of various bar sizes*

Size (mm)	Cross-sectional area (mm²)	Mass per metre (kg)
6[b]	28.3	0.222
8	50.3	0.395
10	78.5	0.616
12	113.1	0.888
16	201.1	1.579
20	314.2	2.466
25	490.9	3.854
32	804.2	6.313
40	1256.6	9.864
50[b]	1963.5	15.413

[a]Derived from the mass per metre run using an assumed density of 0.00785 kg mm^{-2} per metre run. [b]Non-preferred size.

4.3.2 Tolerances

The tolerance applicable to cross-sectional area, expressed as the tolerance on mass per metre length is ±9% for 6 mm, ±6.5% for 8 and 10 mm, and ±4.5% for ≥ 12 mm.

The permitted tolerance on length is ±25 mm.

BS 4466: 1989 specifies the tolerance on bent bars. For a bending dimension up to and including 1000 mm, a tolerance of ±5 mm is permitted. For bending dimensions over 1000 mm and up to and including 2000 mm, discrepancies of +5 mm and −10 mm are acceptable. Over 2000 mm the limits are +5 mm and −25 mm.

For the convenient reinforcement of panels of floor slabs and other similar structural elements reinforcement is available in fabric form. Such fabric is usually produced in square or rectangular meshes of wire or small-size bars welded at their intersections. The range of standard fabric types available from stock is indicated in *Table 4.2*. For prestressing steel the most commonly used sizes of wire, strand and bar are shown in *Table 4.3*.

4.4 Mechanical properties

Some mechanical properties of steels are independent of their microstructures. The elastic moduli of all *plain carbon* structural steels are constant, irrespective of their composition or metallurgical condition (*Tables 4.4* and *4.5*).

All the strength and toughness properties are governed by the steel compositions and their microstructures. Much of the improvement in the available combinations of properties made during the last 20 years or so has been by control of microstructures. The aim is to produce a very low inclusion content and a fine ferrite grain size.

For prestressing steels it is customary for the manufacturer to supply a test certificate defining the size, breaking load,

Table 4.2 Standard fabric types available from stock

Fabric reference	Longitudinal wires			Cross wires			Mass (kg m⁻²)
	Nominal wire size (mm)	Pitch (mm)	Area (mm² m⁻¹)	Nominal wire size (mm)	Pitch (mm)	Area (mm² m⁻¹)	Mass $(kg\,m^{-2})$
Square mesh							
A393	10	200	393	10	200	393	6.16
A252	8	200	252	8	200	252	3.95
A193	7	200	193	7	200	193	3.02
A142	6	200	142	6	200	142	2.22
A98	5	200	98	5	200	98	1.54
Structural mesh							
B1131	12	100	1131	8	200	252	10.9
B785	10	100	785	8	200	252	8.14
B503	8	100	503	8	200	252	5.93
B385	7	100	385	7	200	193	4.53
B283	6	100	283	7	200	193	3.73
B196	5	100	196	7	200	193	3.05
Long mesh							
C785	10	100	785	6	400	70.8	6.72
C636	9	100	636	6	400	70.8	5.55
C503	8	100	503	5	400	49	4.34
C385	7	100	385	5	400	49	3.41
C283	6	100	283	5	400	49	2.61

Wrapping mesh							
D98	5	200	98	5	200	98	1.54
D49	2.5	100	49	2.5	100	49	0.77
Stock sheet size		Length 4.8 m			Width 2.4 m		Area 11.52 m²

Table 4.3   The most commonly[a] used sizes of wire, strand and bar

Type	Nominal diameter (mm)	Specified characteristic breaking load (kN)	Nominal area (mm^2)	Approximate length (m kg^{-1})
Wire	7.0	60.4	38.5	3.31
Strand				
'Standard'	12.0	164	93	1.37
	15.2	232	139	0.92
'Super'	12.9	186	100	1.27
	15.7	265	150	0.85
'Drawn'	12.7	209	112	1.12
	15.2	300	165	0.77
	18.0	380	223	0.57
Bar type/strength				
Smooth or deformed	26.5	568	552	0.230
1030 N mm^{-2}	32	830	804	0.158
	36	1048	1018	0.125
	40	1300	1257	0.101

[a]Products outside British Standard specifications often used are: strand 15.4 mm/ 250 kN and 15.5 mm/261 kN. Refer to manufacturers' publications.

Table 4.4   Physical properties

	Density at 20°C (kg m^{-3})	Coefficient of thermal expansion (10^{-6} K^{-1})	Thermal conductivity (0–100°C) (W m^{-1} K^{-1})	Specific heat (0–100° C) (J kg^{-1} K^{-1})
Pure iron	7870	12.1	78	455
Wrought low carbon steel	7860	12.2	65	480
Cast low carbon steel		12.2	49	
Medium carbon steel	7860	12.2	51	485
High carbon steel	7850	11.1	48	490

Table 4.5   The elastic moduli of steels*

	E (GPa)	G (GPa)	K (GPa)	v
Mild	205–210	80–85	160–170	0.27–0.30
0.75%C	210	80	170	0.29
1.0%C, 1.0%Mn 0.65%Cr, 1.0%W (hardened)	200–205	75–80	165	0.29–0.30
Stainless				
Austenitic	190–205	75–85	130–170	0.25–0.30
Ferritic	200–215	80–85	145–180	0.27–0.30
Martensitic	215	85	165	0.28

* $G = \dfrac{E}{2(1+v)}$; $K = \dfrac{E}{3(1-2v)}$.

These data (taken from a number of sources) are subject to measurement errors of, mostly, about ±5%.

relaxation type and load-extension properties (the latter covering at least elongation at fracture), 0.1% proof load or load at 1% extension and modulus of elasticity. These and other characteristics are illustrated in *Figure 4.4*.

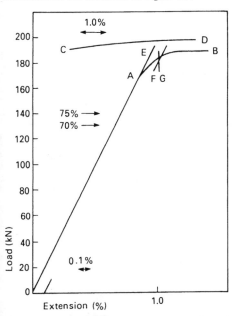

Figure 4.4 *Representative load-extension diagram for a 12.9 mm strand. OA, elastic range; A, limit of proportionality; C, continuation of B at lower extension magnification; D, elongation at fracture (maximum load); OE, modulus line or line of proportionality; F, 0.1% offset from OE; G, 1.0% total extension; 70% and 75%, indication of prestressing load range*

The requirements for relaxation and typical examples are given in *Tables 4.6* and *4.7*.

Many factors influence the properties of steel. These include:

- Chemical composition
- Effect of cold working

Figures 4.5 and *4.6* illustrate, respectively, the effect of sulphur content on toughness and the effect on tensile properties of cold working.

4.4.1 Weldable structural steels

BS 4360: 1986 was until recently the cornerstone in Britain defining structural steels which may readily be welded. In 1990, this was replaced by two new standards: BS EN 10 025:

Table 4.6   *Maximum 1000 h relaxation[a] at 20°C*

	Initial load/breaking load (%)		
	60	70	80
Bar	3.5	6.0	
Strand and straight wire			
Relaxation class 1	4.5	8.0	12.0
Relaxation class 2	1.0	2.5	4.5

[a]From BS 5896: 1980 and BS 4486: 1988.

Table 4.7   *Typical 1000-h relaxation at 20°C for strand*

Initial load/breaking load (%)	Relaxation 1 (%)	Relaxation 2 (%)
60	3.4	1.00
65	4.4	1.04
70	5.6	1.10
75	7.6	1.60

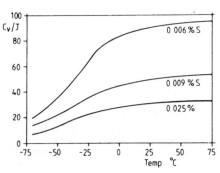

Figure 4.5   *The effect of sulphur content on the toughness of high-strength low-alloy (HSLA) steel (minimum yield stress $R_s = 450\,N\,mm^{-22}$). (Reproduced by kind permission of the Director, The Steel Construction Institute)*

1990. Hot rolled products of non-alloy structural steels and their technical delivery conditions, and BS 4360: 1990 Weldable structural steels (it is intended that this too will be replaced by a series of four new European standards).

For the convenience of those who need to make comparisons the equivalences between BS 4360: 1986 and the 1990 standards are shown in *Table 4.8*. A rough guide to the cost differences for different grades of steel appears in *Table 4.9*. The limits on permissible carbon equivalent are shown in *Table 4.10*.

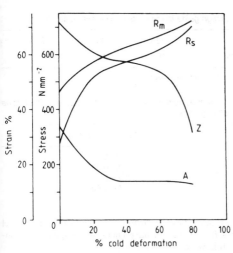

Figure 4.6 *The effects of cold work on the tensile properties of steel (0.13 0.18%C, 0.6–0.9%Mn). (Reproduced by kind permission of the Director, The Steel Construction Institute)*

Table 4.8 *Equivalences between the two standards published in 1990 and the 1986 version of BS 4360*

BS 4360: 1986	BS EN 10 025: 1990
40B	Fe 360 B
40C	Fe 360 C
40D	Fe 360 D1, Fe 360 D2
43B	Fe 430 B
43C	Fe 430 C
43D	Fe 430 D1, Fe 430 D2
50B	Fe 510 B
50C	Fe 510 C
50D	Fe 510 D1, Fe 510 D2
50DD	Fe 510 DD1, Fe 510 DD2

BS 4360	
1986	*1990**
40E	40EE
43E	43EE
50E	50EE
50F	50F
55C	55C
55E	55EE
55F	55F

*Nomenclature is unchanged but grades E have been upgraded to EE.

Table 4.9   *Rough guide to the costs of different grade steels*

Grade	43A	50B	55C
Relative cost*	100	110	125

*100, cheapest; 125, most expensive.

Table 4.10   *The maximum permissible carbon equivalent (CE) value for different steel grades*

Grade	Maximum CE (%)
40	0.39–0.41
43	0.39–0.41
50	0.43–0.47
55	0.41–0.53

Weldability is limited by the carbon equivalent value of the steel (see *Table 4.10*). Steels will have chemical compositions containing the following:

- Carbon 0.16–0.25%
- Manganese 1.5–16%
- Silicon 0.1–0.5%.

The higher grades (50 and 55) will also include small quantities of aluminium, niobium and vanadium. Tensile and impact properties vary a little with product form and thickness. Typical data for flat products are shown in *Table 4.11*.

Weathering grade equivalents are available for grades 50A, 50B and 50C. These are designated WR50A, etc., and have extra additions of chromium, copper and, in 50A, nickel, which confer the weathering quality. The mechanical properties of these steels differ little from those of the equivalent non-weathering grades.

4.4.2 Welding of steels
Welding is a complex operation which involves not only the formation of a joint but also thermal cycling and chemical reactions. Molten weld metal must be protected from adverse reactions with the environment. Differential expansions and contractions can cause distortions and/or residual stresses and the local heat treatment of the steels can cause microstructural and property changes.

Structural steels, as defined by BS 4360: 1986, are chosen for their weldability, i.e. so that property changes brought about by welding produce a joint which is acceptable for the service envisaged.

When the weld zone is raised to the melting temperature of the steel, the solid metal immediately adjacent to it is heated to temperatures well within the austenite range. On removal of the

Table 4.11 *Properties of structural steels according partly to BS 4360: 1986 (plates and wide flats)*

Grade	Tensile strength (N mm^{-2})	Minimum yield strength at 16 mm (N mm^{-2})	Charp V-notch impacts 27 J at (°C)
40A	340/500	235	–
40B	340/500	235	20†
40C	340/500	235	0
40D	340/500	235	–20
40EE	340/500	260	–50
43A	430/580	275	–
43B	430/580	275	20†
43C	430/580	275	0
43D	430/580	275	–20
43EE	430/580	275	–50
50A	490/640	355	–
50B	490/640	355	20†
50C	490/640	355	0
50D	490/640	355	–20
50DD	490/640	355	–30
50EE	490/640	355	–50
50F	490/640	390	–60
55C	550/700	450	0
55EE	550/700	450	–50
55F	550/700	450	–60

†The specified impact values are verified by test only at the request of the purchaser.

heat source, the whole reaction cools at rates determined by the conduction of heat into the surrounding cold metal. These rates of cooling can be very rapid, sometimes exceeding 1000 C s^{-1}.

Depending on the steel composition and the rate of cooling, the local microstructure might revert to ferrite and pearlite or might form martensite or bainite. The properties of these products are shown in *Figure 4.7*. To avoid problems such as quench cracking, hydrogen embrittlement and the formation of locally brittle regions, a maximum hardness of about HV = 350 kg mm^{-2} is taken as an acceptable rule of thumb (tensile test data corresponding to this hardness are given in *Figure 4.7*).

Ways in which suitable weld properties may be obtained are:

(1) By choosing appropriate material to be welded
(2) By choosing a suitable welding method (energy input) and consumables
(3) By preheating the weld zone and surrounding metal.

Welds may be of fillet or butt configuration; these are illustrated in *Figure 4.8*. Welding should only be performed by fully qualified welders. However, defects may occur, some of

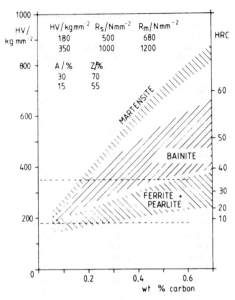

Figure 4.7 *The effect of carbon content on the hardness of various steel microstructures. (Reproduced by kind permission of the Director, The Steel Construction Institute)*

which are shown in *Figure 4.9*. The theoretical strength of structural welds should be determined by a qualified structural engineer. Guidance in this exercise may be obtained using the load capacities indicated in *Tables 4.12*, *4.13* and *Figure 4.10*.

Table 4.12   *Static load capacities of full penetration butt welds**

Throat thickness (mm)	Shear load capacity $(kN\,mm^{-1})$		Tension or compression load capacity $(kN\,mm^{-1})$	
	Grade 43	Grade 50	Grade 43	Grade 50
5	0.83	1.06	1.38	1.78
10	1.65	2.13	2.75	3.55
20	3.18	4.08	5.3	6.8
40	6.36	8.16	10.6	13.6

*Data taken from 'Steelwork Design: Guide to BS 5950, Part 1, 1985, volume 1. Section Properties and Member Capacities', second edition (1987), Steel Construction Institute.

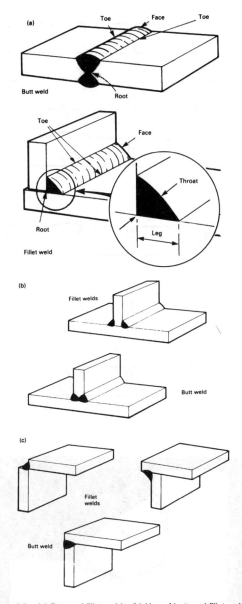

Figure 4.8 (a) Butt and fillet welds. (b) Use of butt and fillet welds to make T joints. (c) Use of butt and fillet welds to make corner joints. (Reproduced from A Guide to Designing Welds with kind permission of Abington Publishing)

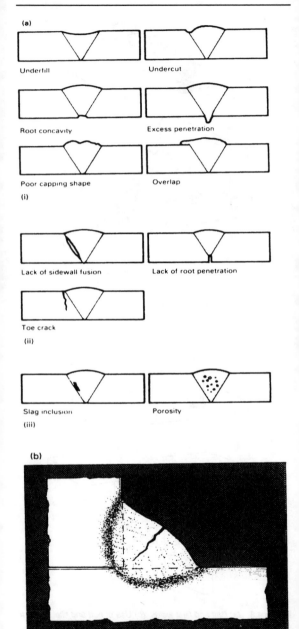

(a)

Underfill

Undercut

Root concavity

Excess penetration

Poor capping shape

Overlap

(i)

Lack of sidewall fusion

Lack of root penetration

Toe crack

(ii)

Slag inclusion

Porosity

(iii)

(b)

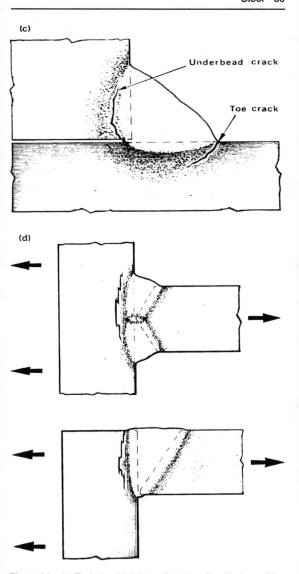

Figure 4.9 *(a) Typical weld defects: (i) weld profile; (ii) planar; (iii) volumetric defects. (Reproduced from* A Guide to Designing Welds *with kind permission of Abington Publishing.) (b) Transverse hot crack (solidification cracking). (c) Cracks in heat-affected zone. These are often made worse by high local hydrogen contents. (d) Lamellar tearing. (Reproduced from* Weldability of Steel *with kind permission of The Welding Institute)*

Table 4.13 *Static load capacities of fillet welds between plates at right angles**

Leg length† (mm)	Load capacity (kN mm^{-1})	
	Grade 43 electrode E43 weld $\sigma_y = 215\,N\,mm^{-2}$	Grade 50 electrode E51 weld $\sigma_y = 255\,N\,mm^{-2}$
5	0.753	0.893
10	1.51	1.79
15	2.26	2.68
20	3.01	3.57

*Data taken from 'Steelwork Design: Guide to BS 5950, Part 1, 1985, volume 1, Section Properties and Member Capacities', second edition (1987), Steel Construction Institute.
†Throat thickness = 0.7× leg length.

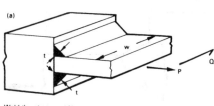

(a)

Weld throat area $= t \times w$
Weld stress for tension load $P = P/2tw$
Weld stress for shear load $Q = Q/2tw$

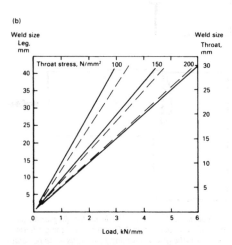

(b)

Figure 4.10 *(a) Calculation of the stress on a fillet weld. (b) Fillet weld design chart showing the strength of fillet welds in terms of load/mm. (Reproduced from* A Guide to Designing Welds *with kind permission of Abington Publishing)*

4.5 Other steels

This chapter has so far dealt with steels for structural use including those for reinforcing concrete. There are other steels for general engineering use which are also covered by BS 1449: Part 1: 1983. A simplified abstract of the mechanical properties of these steels is indicated in *Table 4.14*.

4.5.1 Stainless steels
The quality of stainlessness is imparted to steels by the presence of at least 12% of chromium in the steel. Oxidation and/or corrosion produces a dense adherent oxide film which acts as a barrier to further corrosion.

The presence of large amounts of chromium in the steel has profound effects on its metallurgy. This is further compounded by the presence of further alloying elements.

Broadly, there are three types of stainless steel: martensitic, ferritic, and austenitic. Martensitic steels contain, typically, approximately 13% chromium and more than 0.1% carbon. Ferritic steels contain very little carbon (< 0.08%) with chromium contents varying from 13% to 17%. Austenitic steels contain a minimum of 8% nickel in addition to the chromium. Carbon contents must be maintained as low as practicable, typically below 0.1%. The relevant British Standard is BS 1449: Part 2: 1983; examples of mechanical properties are given in *Table 4.15*.

4.5.2 Hollow sections
Tubes are available in the full range of carbon, carbon–manganese and stainless steels. They may be made by seamless tube making methods or by methods which involve a welded seam. Finishing operations may be carried out hot or cold and the finished tubes may or may not be heat treated.

BS 6323: 1982 deals with a range of tubes for general engineering purposes.

Table 4.16(a) indicates the method of manufacture and designation of some tubes. *Table 4.16(b)* gives an indication of grades and ultimate tensile strengths for some tubes.

4.5.3 Steels for fasteners
4.5.3.1 Black bolts and nuts Black bolts and nuts are usually fashioned by finish machining of hot- or cold-forged steel. The composition of the steel is left to the discretion of the manufacturer except for controls on nitrogen, sulphur and phosphorus contents. For nuts, the requirements are less exacting than for bolts.

The relevant standard is BS 4190: 1967, which grades bolts and nuts by their mechanical properties. It follows from ISO recommendations.

The grading system for bolts uses a two-number code, e.g. 4.8. The first number is the minimum UTS/10, where the ultimate tensile strength (UTS) is in kilogram force per

Table 4.14 Mechanical properties of carbon–manganese and microalloyed steels*

Processing	Grade	σ_y (MPa)	UTS (MPa)	Minimum percentage elongation		Mandrel diameter for 180° bend†	BS 4360: 1986 equivalent‡
				50 mm	80 mm		
Carbon–manganese steels							
H or C	34/20	200	340	29	27	2a	
H or C	37/23	230	370	28	26	2a	40B
H only	43/25	250	430	25	23	3a	43A
H only	50/35	350	500	20	18	3a	50B
Microalloyed steels§							
H or C	40/30	300	400	26	24	2a	
H or C	43/35	350	430	23	21	2a	
H or C	46/40	400	460	20	18	3a	
H or C	50/45	450	500	20	18	3a	
H or C	60/55	550	600	17	15	3.5a	55C

*Based partly on BS 1449: 1984; courtesy of British Standard Institution. For qualifying remarks, refer to the standard.
†a = sheet thickness.
‡Equivalences to BS 4360: 1986 are not exact.
§Microalloyed grades are also available with improved formability; the percentage elongation is c. 2% more, the mandrel diameter in the bend test c. 2a less, whilst the strengths are the same; these grades are designated 40F30, etc.

Table 4.15 Compositions and properties of stainless steels (all in softened condition)*

Grade	Typical composition (%)				0.2% proof stress (MPa)	Ultimate strength (MPa)	Elongation (%)	Sensitization time† (min)
	C	Cr	Ni	Other				
Ferritic								
403 S17	0.08	13			245	420	20	
430 S17	0.08	17			245	430	20	15
Austenitic								
304 S15	0.05	18	9		195	500	40	0
309 S24	0.12	23	14		205	510	40	0
310 S24	0.12	24	20		205	510	40	
316 S11	0.03	17	12	2.25 Mo	190	490	40	30
317 S12	0.03	18	15	3.5 Mo	195	490	40	30
320 S31	0.08	17	12	2.25 Mo Ti = 5C	210	510	40	30

*Based partly on BS 1449: Part 2: 1983; courtesy of British Standards Institution.
†Time quoted is the time for which a sample is held at 650°C to bring about sensitivity to stress corrosion cracking: this is assessed according to BS 5903 (see Sections 5.10.8 and 5.10.9).

Table 4.16   *Steel tubes**
(a)

Method of manufacture	Designation
Hot finished, welded	HFW
Hot finished, seamless	HFS
Cold finished, seamless	CFS
Electric resistance welded (including induction welded)	ERW
Cold finished electric resistance welded (including induction welded)	CEW
Submerged arc welded	SAW
Longitudinally welded stainless	LW
Cold finished longitudinally welded stainless	LWCF

(b)

Type and grade of steel	Minimum UTS (MPa)	Grade
Carbon and carbon–manganese		
0.2 C max.	360	3
0.25 C max.	410	4
0.2–0.3 C, 1.2–1.5 Mn	650	7
Austenitic stainless		
304 S15	510	21
316 S13	480	22
321 S31	510	24

*Based partly on BS 6323: 1982; courtesy of British Standards Institution.

Table 4.17   *Properties of steels for black bolts and nuts**

Property	Grade		
	4.6 hot forged	*4.8 cold forged*	*6.9 cold forged*
Yield stress (MPa)	235	314	530
UTS (MPa)	392	392	588
Yield stress/UTS	0.6	0.8	0.9
Elongation to fracture (%)	25	14	12
Hardness, HV (30)	110–170	110–170	170–245

*Based partly on BS 4190: 1967; courtesy British Standards Institution.

square millimetre $(kgf\,mm^{-2})$. The second number is $10 \times$ (yield stress/UTS).

The details of properties and grades are given in *Table 4.17*. For nuts, only the first number is used.

4.5.3.2 High strength friction grip bolts   These are manufactured by forging, after which the steel is hardened by quenching and tempering. No specifications are laid down for composition, except for maximum limits on nitrogen, sulphur and phosphorus contents. The bolts are specified according to their load-carrying ability. The material properties taken for this purpose are listed in *Table 4.18*.

The relevant standard for the bolts is BS 4395: 1969.

Table 4.18 *Properties of steels for high strength friction grip bolts*[*†]

Range of thread diameter (mm)	Yield or 0.2% proof stress (MPa)	UTS (MPa)	HV 30 (kg mm^{-2})
General range			
12–24	635	827	260–330
27–36	558	725	225–292
Higher grade bolts			
16–33	882	981	280–380

*Based partly on BS 4395: 1969; courtesy British Standards Institution.
†Further tests are mandatory for full compliance with the standard. These include measuring the load required to cause the bolt length to extend by 12.5 μm and tests for strength under conditions which simulate a wedge under the bolt head.

4.5.4 Ropes and roping steel

Steel wires for ropes are required to have high tensile strength. They contain between 0.5 and 1.0% carbon and are cold drawn. A process known as 'patenting', involving working and annealing at temperatures which do not produce austenite, imparts sufficient ductility. The resulting microstructures contain very small spherical particles of cementite in a work-hardened ferrite matrix.

Strength grades are defined in BS 2763: 1982 (which incorporates recommendations given in ISO 2232: 1973 and ISO 3154: 1976). A twisting ductility and a resistance to reversals of bending are also required. Examples are given in *Table 4.19*.

The wires are available with a bright, drawn finish or in several zinc coated versions. Variations exist in the size ranges and finishes which can match an individual strength.

Details are given in the standard.

4.5.5 Steel castings

A wide variety of complex shapes may be made in the form of steel castings. In the as-cast condition, the microstructures are variable from surfaces to bulk, there may be more porosity and residual stresses. Heat-treatment gives more uniform structures and anneals out residual stresses. Porosity is usually not affected by heat treatment but, as long as its amount is not too large, and it is not in critical regions of the casting, it may be acceptable. Inspection by testing for pressure tightness and by radiography and ultrasonic methods may be advisable, depending on the proposed use.

The relevant standard for general engineering castings is BS 3100: 1991; castings for containing pressure are specified in BS

Table 4.19 *Strength and fracture requirements for general-purpose and high-duty rope wire**

Strength grade (MPa)	Minimum No. of twists to failure for a wire length of 100 diameters		Minimum No. of reverse bends through a radius of 3.75 mm (general-purpose and high-duty wire)
	General-purpose wire	High-duty wire	
1370	31	NA	14
1570	30	34	14
1770	27	31	13
1860	25	28	–
1960	23	25	12
2050	22	NA	10
2150	20	NA	10

NA, not available.
*Data are for bright drawn wire, 1.0 1.3 mm diameter. For other finishes and diameters, data will be different. For details refer to BS 2763: 1982.

1504: 1976. The 1976 version of BS 3100 uses a coding system which replaces a variety of earlier standards. The code consists of one or two letters followed by a number (*Table 4.20*). The number is arbitrary. There are different rules for stainless steels.

The compositions and mechanical properties of some selected steels are given in *Table 4.21*. Data for stainless steels are not given here; for details see BS 3100: 1991.

4.6 Fire resistance

The Building Regulations (Part E) demand that the load-bearing members of a structure should not fail before a preselected time has elapsed. Test details are given in BS 476: Part 8: 1972. For simply supported beams with distributed load, failure is defined as a vertical deflection at centre span of $L/30$, where L is the span. Loaded beams are heated at a standard rate until failure occurs and the time to failure is measured.

Table 4.20 *Letter codes for steel castings**

First letter (steel type)	
A	Carbon or carbon–manganese steels
B	Low alloy steels
Second letter (use)	
None	General-purpose use
L	Low-temperature toughness
W	Wear resistance
T	High tensile strength
M	Magnetic properties

*Reproduced partly from BS 3100: 1991; courtesy of British Standards Institution.

Table 4.21 Carbon and carbon–manganese steel castings*

Constituent	Grade						
	A1	A2	A3	A4	A5	A6	AL1
Carbon (%)	0.25	0.35	0.45	0.25	0.33	0.33	0.20
Silicon (%)	0.60	0.60	0.60	0.60	0.60	0.60	0.60
Manganese (%)	0.90	1.0	1.0	1.60	1.60	1.60	1.10
Phosphorus (%)	0.05	0.05	0.05	0.05	0.05	0.05	0.04
Sulphur (%)	0.05	0.05	0.05	0.05	0.05	0.05	0.04
Residuals (%)	–†	–	–	–	–	–	–
Heat treatment‡	At manufacturer's discretion	–	–	N N+T H+T	N N+T H+T	H+T	At manufacturer's discretion
Max. section (mm)	–	–	–	–	100	63	–
Lower yield stress (MPa)	230	260	295	320	370	495	230
UTS (MPa)	430	490	540	540	620	690	430
Elongation on 5.65 $A_0^{1/2}$ (%)	22	18	14	16	13	13	22
Charpy V-notch test T(°C)	20	20	20	20	20	20	–40
Energy (J)	27	20	18	30	25	25	20

*Reproduced partly from BS 3100: 1991; courtesy British Standards Institution.
†CR <0.25%, MO <0.4%, Ni <0.4%, Cu <0.3%, total <0.8%.
‡N, normalized; H, heat-treated by quenching into oil or water; T, tempered.

Fire resistance times of 0.5, 1, 1.5, 2, 3 or 4 h are required depending on the nature, use and contents of the building.

To test large beams is costly and only a small number of full-scale tests have been done. Not unexpectedly, these tests show that to reach the specified failure deflection the lower flange must reach temperatures which reduce as the design stress is increased. The rate of temperature rise is a function of the sectional shape, specifically of its perimeter/area ratio (*Figure 4.11*).

Where structural steelwork requires protection to provide fire resistance this may be in one of the following forms:

- Cladding with concrete or other materials such as vermiculite (see Chapter 20)
- Intumescent paint (see Chapter 14).

A relatively new sub-discipline known as fire engineering may provide solutions which do not require extensive casing of the steel. For further information on this technique consult BRE.

4.7 Corrosion and corrosion protection

Inevitably, steel will rust when in the presence of moisture and oxygen. About 1 tonne of steel is lost in the UK every 90 s. The rate of attack varies substantially, depending on the environmental conditions and the methods used to protect the metal. Careful design can do much to reduce corrosion rates.

Three main types of corrosion are considered here.

4.7.1 Uniform corrosion

The rate of uniform attack is a function of the environment and the material (*Table 4.22*). In clean rural environments, a relative humidity of about 60% or more is required to cause noticeable rusting (*Figure 4.12*). The rust forms a layered structure on the surface, consisting of various iron oxides and hydroxides. They are generally impure and can retain moisture and water-soluble salts, which can accelerate further corrosion. In atmospheres containing chlorides or sulphur oxides, the rust can concentrate these materials, generating pits which can be both dangerous to performance and difficult to remove.

Weathering grades of steel are designed to produce oxide layers of different character. Chromium and copper in the steel help to produce a dense and adherent oxide film which acts as a barrier to the environment. Corrosive substances can still pass through the film but at a rate which reduces as the film thickens. As a result, corrosion rates decrease parabolically with time (*Figure 4.13*). The film is protective only if it remains intact. Mechanical or chemical disruption allows corrosion to continue. It follows that when using weathering steels care must be taken to avoid such disruption.

(a)

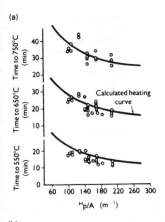

(b)

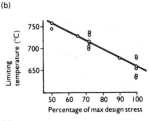

(c)

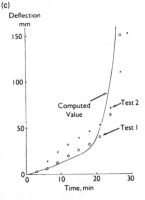

Figure 4.11 *(a) Composition of observed and predicted data on the effect of the section factor on the time taken for the lower flange to reach various temperatures in a BS 476: Part 8: 1972 test. (b) Effects of stress on the limiting temperature for a deflection of L/30 in a BS 476: Part 8: 1972 fire test. (c) Comparison of observed and predicted deflection vs. time curves for an unprotected, fully loaded 356 mm × 171 mm, 67 kg m⁻¹ steel beam in a BS 476: Part 8: 1972 fire test. (Reproduced from Metals and Materials, the journal of the Institute of Metals, Vol. 2, No. 1, 1986, p. 25, and Vol. 2, No. 5, 1986, p. 274)*

Table 4.22   *Examples of metal loss rates* (μm year $^{-1}$)

	Environment		
	Rural	*Maritime*	*Industrial*
Mild-steel to BS 4360: 1986	5	6.5	15–75
Weathering steel to BS 4360: 1986	1.3	4	2.5

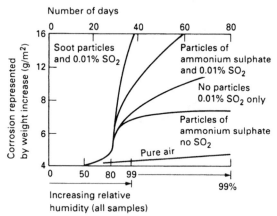

Figure 4.12   *The influence of relative humidity and pollution on the corrosion of iron. (Reproduced from* Controlling Corrosion. 1. Methods, *with permission of HMSO)*

4.7.2 Galvanic corrosion

This is the most common type of corrosion. Electrochemical cells are set up on the surface by contact with other metals or by local variations in the metal itself or in the environment.

A simple electrochemical corrosion cell is shown in *Figure 4.14*. Two different metals are immersed in an electrolyte. These metals acquire different electrode potentials and when connected through an external circuit a current flows. The anodic electrode corrodes and the cathodic electrode is protected as a consequence of ionic drifts in the electrolyte to close the circuit.

Anodes and cathodes arise in many ways. On a plain metal surface, some grains are anodes and others cathodes; grain boundaries are often anodic to the grains either side of them. Regions which have been cold worked are anodic to regions which have not been; nail heads are anodic to nail shanks. Rust and mill scale is cathodic to bare steel; this is why it is important to apply paint over clean and dry surfaces.

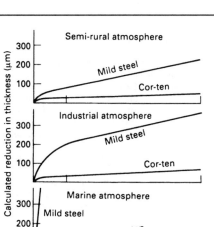

Figure 4.13 *The effects of small alloying additions on the resistance of steel to atmospheric corrosion. Cor-Ten contains 2.3% alloying elements, particularly copper, chromium and phosphorus.*
(Reproduced from Controlling Corrosion. 1. Methods, *with permission of HMSO)*

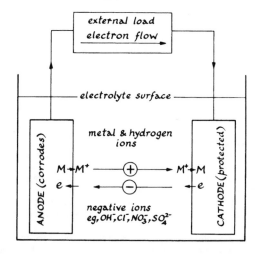

Figure 4.14 *A galvanic corrosion cell. Internally, the cathode receives electrons from the external circuit. Externally the cathode receives electrons from the corrosion cell. To stop corrosion one of the following must be done: (i) stop the anode reactions; (ii) stop the cathode reactions; and/or (iii) stop the electron flow in the external circuit*

Bimetallic corrosion can be particularly severe and often arises at design details where different materials are used. Metals can be arranged in an electrochemical series (*Table 4.23*). The metals at the base end will be anodes when in contact with metals from the noble end. The further apart are the metals in the series, the greater is the difference in electrode potentials and the more aggressive is the attack on the anode. Thus, when iron is in contact with copper, the iron corrodes at an accelerated rate and the copper is protected. When iron and zinc are in contact, the iron is cathodic and protected whilst the zinc is anodic and corrodes. This is the principle behind the use of galvanizing.

Note that stainless steels are listed twice in *Table 4.25*. When covered with an intact oxide film, they are passive, but when in conditions which disrupt the oxide film they are exposed to the environment and can corrode, i.e. they are active.

Table 4.23   *The electrochemical series*

Base	Magnesium
	Zinc
	Aluminium (commercially pure)
	Cadmium
	Duralumin
	Mild-steel
	Cast-iron
	Stainless steels (active)
	Lead
	Tin
	Nickel
	Coper alloys
	Stainless steel (passive)
	Silver
	Graphite
	Gold
Noble	Platinum

4.7.3 Crevice corrosion

Crevices arise in many ways, through bad design details or by the generation of pits. The significant thing about crevices is that they are sites starved of oxygen and in which aggressive substances from the environment can be concentrated. Even without the impurities, oxygen starvation can accelerate corrosion in the crevice if the rest of the structure is in contact with a good oxygen supply.

Figure 4.15 shows an electrochemical cell composed of two pieces of the same steel immersed in electrolytes which are in contact with one another and which differ only in their oxygen content. One electrolyte is enriched in oxygen and the other is starved of oxygen. A potential difference arises, the oxygen starved electrode being the corroding anode.

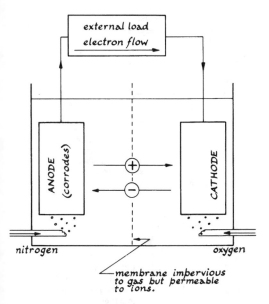

Figure 4.15 *A differential aeration cell*

Crevice corrosion is particularly insidious because the attack is hidden from view in the crevice. Examples of common crevices are shown in *Figure 4.16*. The best way to avoid crevice corrosion is to design away the crevices.

Differential aeration can arise in ways other than at crevices. For example, should a pipeline pass through regions of sandy soil and clay, ready oxygen access in the sandy regions makes the pipe cathodic and corrosion is encouraged in the oxygen impoverished clay. Another example is given in *Figure 4.17* which shows the same principles operating under a single drop of water.

4.7.4 Splash zones
Parts of structures which are immersed in water but are in the zone which alternately dries and is re-wetted are particularly susceptible to corrosion. The main reason for this is that during the drying phase the remaining water droplets become more concentrated in aggressive salts. The corrosion rate as a function of height in relation to water levels is shown in *Figure 4.18*.

4.7.5 Corrosion protection
For corrosion to occur, there must be liquid water in contact with the metal. Designs which shed water are desirable. It should also be remembered that accumulations of dirt and

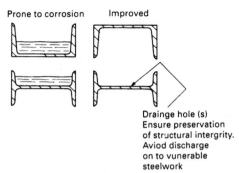

(a) Retention of dirt and water

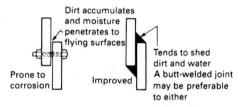

(b) Comparison of bolted and welded lap joints

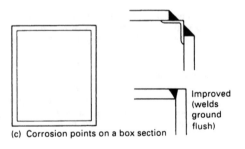

(c) Corrosion points on a box section

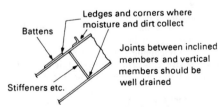

(d) Corrosion points on inclined members

Figure 4.16 *Corrosion points*

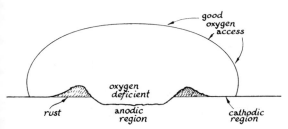

Figure 4.17 *Differential aeration due to a single drop of water*

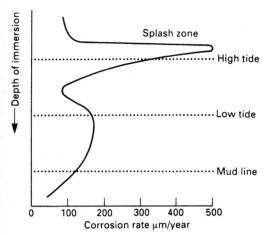

Figure 4.18 *Corrosion rates at different positions for a steel structure partly immersed in sea water. (Reproduced from* Corrosion Control in Engineering Design *with permission of the DOI)*

other debris can hold water for surprisingly long times. Simple changes which introduce drainage holes and which avoid introducing narrow gaps, ledges, etc., can do much to reduce corrosion. Access for cleaning and painting is also desirable. Extensive recommendations can be found in BS 5493: 1977.

The avoidance of galvanic corrosion also begins with good design and choice of materials. Steel fasteners are often used in aluminium structures because aluminium fasteners are not strong enough. To avoid galvanic corrosion, it is necessary to open the galvanic circuit; if no current flows, there is no corrosion. This can be achieved using electrically insulating sleeves and gaskets (*Figure 4.19*). Alternatively, thought might be given to replacing a fastened joint by a welded one.

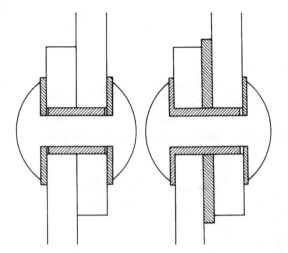

Figure 4.19 *Insulated joints for use when plates and fasteners are electrochemically different*

Welded joints can be sites of local corrosion. This might arise from small differences in electrode potentials between the base and weld metal, but is more likely to do with the position and surface finish of the weld. Rough weld surfaces and welds in corners are more likely to trap moisture and dirt than are smooth welds away from corners.

4.7.6 Surface treatment

Coatings fall broadly into those which provide active protection through electrochemical action and those which are inert barriers to the environment. Whichever type of coating is used, it is important that it be applied to a properly prepared surface.

Coatings may take many forms, some of which are indicated below:

- Zinc (see *Figure 4.20*)
- Aluminium
- Cadmium
- Paints (see Chapter 14 and *Table 4.24*)
- Cement
- Phosphating
- Chromating
- Bitumen (see Chapter 7)
- Polymers (see Chapter 16 and *Table 4.25*)

Cathodic protection may also be used to resist corrosion. Information on this specialized technique may be obtained from the HMSO publication *Controlling Corrosion 1.*

Table 4.24 *Main generic types of paint and their properties**

Paint type	Cost	Tolerance of poor surface preparation†	Chemical resistance	Solvent resistance‡	'Over-coatability' after ageing§	Comments
Bituminous	Low	Good	Moderate	Poor	Good—with coating of same type	Limited to black and dark colours. Thermoplastic
Oil based	Low	Good	Poor	Poor	Good	Good decorative properties
Alkyd, epoxy ester, etc.	Low/medium	Moderate	Poor	Poor/moderate	Good	
Chlorinated rubber	Medium	Poor	Good	Poor	Good	High-build films remain soft and are susceptible to 'sticking' during transport
Vinyl	High	Poor	Good	Poor	Good	
Epoxy	Medium/high	Very poor	Very good	Good	Poor	Susceptible to 'chalking' in ultraviolet light
Urethane	High	Very poor	Very good	Good	Poor	Better decorative properties than epoxies
Inorganic silicate	High	Very poor	Moderate	Good	Moderate	May require special surface preparation

*Reproduced by kind permission of the Director, The Steel Construction Institute.
†Types rated poor or very poor should only be used on blast-cleaned surfaces.
‡Types rated poor or very poor should not generally be overcoated with any other type.
§Types rated poor or very poor require suitable preparation of the aged surface if they are to be overcoated after an extended period.

Table 4.25   *Summary of properties of weathering and top coats used on organic coated steel systems for building cladding*

	Advantages	Limitations
Plastisol	A thick tough leather grain embossed coating. Excellent range of properties. Withstands site handling	Some limitations on use outside UK
Pvf_2	Good colour retention and durability. Excellent resistance to fats and oils, and to petroleum oils and hydrocarbons	Coating more easily scratched than Plastisol. Poor mark resistance
Architectural polyester	Good exterior durability, flexibility and temperature stability	Medium-life product. Exhibits slight chalking and colour changes in high ultraviolet environments outside UK
Silicone polyester	Good exterior durability. Hard stain-resistant coating. Good temperature stability	Low flexibility

*Reproduced from *Metals and Materials*, the journal of the Institute of Metals, Vol 4, No. 6. 1988. p. 359.

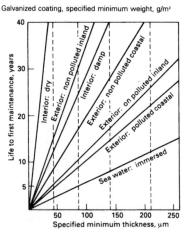

Galvanized coating, specified minimum weight, g/m²

Figure 4.20   *Typical lives of zinc coatings in selected environments. (Reproduced partly from BS 5493: 1977; courtesy of British Standards Institution)*

Further methods of combatting corrosion involve the use of weakening steel and stainless steels. For further information see Section 4.5.

Bibliography

MOON, J.R. In DORAN, D.K. (ed.), *Construction Materials Reference Book*, Butterworth-Heinemann, Oxford, Chapter 5 (1992)

5 *Other metals*

The metals included in this chapter are gold, silver, copper, nickel, lead, tin and zinc. The first three can be classed as noble metals in the sense that they lie above hydrogen in the electrochemical classification of metals (i.e. they cannot release hydrogen from solution) and hence exhibit good resistance to corrosion by acids. Nickel, also, is passive in neutral and alkaline solutions. With the exception of zinc, however, all these metals exhibit a corrosion resistance that is far superior to that of unalloyed iron or aluminium, and numerous archaeological artefacts testify to the outstanding durability of these metals. The last three, lead, tin and zinc, together with cadmium, are usually referred to as 'white metals' since they are the main constituents of the soft, metallic bearings which have a whitish colouration.

It should be appreciated that none of these metallic elements find tonnage application in the main structures of buildings. However, in each case they have important combinations of properties which are exploited in the construction industry.

Table 5.1 lists some data for the production and the reserves of the metals considered. Data on aluminium and iron are given for comparison. Column 1 lists the average concentration of the metals in the earth's crust, smaller concentrations are also present in seawater. On the basis of these data, one might expect gold to be expensive; but, on the same basis, one would expect lead to be more expensive than copper while zinc should be priced similarly to nickel. The limited data on the scarcity value, indicated by the estimated static lifetime of the known workable deposits, also bears no relation to the cost of the metal. The last column of *Table 5.1* shows the level of concentration of the metal, relative to the average concentration, which is required to produce a workable ore body.

General physical and mechanical properties

Table 5.2 lists data, taken mainly from ASTM sources, of typical values of properties of the pure metals. These data should be used with caution, since properties can change markedly when small amounts of impurities or deliberate trace additions are present.

5.1 Copper

Copper has been used for over 9000 years. It is very ductile with comparatively good strength and toughness; it is non-magnetic and has a higher electrical and thermal conductivity than any of

Table 5.1 Metal resources in the earth's continental crust

Metal	Average concentration (g t⁻¹)	Mine production 1986 (Mt)	Recoverable reserves (Mt)	Static lifetime (years)	Workable concentration / Average concentration
Copper	55	8.11	1000	120	73
Gold	0.004				250
Lead	13	3.4	170	50	3000
Nickel	75	0.73	1200	1650	2500
Tin	2	0.17	25	150	570
Zinc	70				
Aluminium	82 300				4
Iron	56 300				4

Table 5.2 Physical and mechanical properties of the pure metals

Property	Metal						
	Copper	Gold	Nickel	Silver	Lead	Tin	Zinc
Atomic number	29	79	28	47	82	50	30
Atomic weight	63.54	196.97	58.71	107.89	207.21	118.69	65.38
Atomic radius (nm)[5]	12.75	14.41	12.45	14.44	17.49	15.1	13.3
Crystal structure	f.c.c.	f.c.c.	f.c.c.	f.c.c.	f.c.c.	b.c.t.	c.p.h.
Melting point (°C)[5]	1083	1063	1453	960.8	327.4	231.9	419.5
Boiling point (°C)[5]	2595	2970	2730	2210	1725	2270	906
Density (g cm^{-3})[5]	8.96	19.32	8.9	10.49	11.34	7.30	7.14
Specific heat at 20°C (J g^{-1})[5]	0.380	0.131	0.440	0.234	0.129	0.230	0.383
Thermal conductivity 0–100°C (J cm^{-2} s^{-1} C^{-1})	3.93	2.97	0.92	4.19	0.347	0.377	1.13
Coefficient of linear thermal expansion[2] 0–100°C (cm cm^{-1} C^{-1}) × 10^6	16.5	14.2	13.3	19.68	29.3	23	39.7
Electrical resistivity at 20°C (μΩ cm^{-1})[5]	1.673	2.35	6.84	1.59	20.65	13	5.92
Temperature coefficient of electrical resistance (°C^{-1})	0.0039	0.004		0.0041	0.0042	0.00447	0.00419
Yield or 0.2% proof stress (N mm^{-2}) Annealed	60	0	103	55	8	14	
Hard	325	205	620				
Ultimate tensile stress (N mm^{-2}) Annealed	220	130	410	130	16	16	140

	385	220	700		40	90	160
Hard							
Elongation (%)							
Annealed	55	45	45	54			60
Hard	4	4	15	48	4	5	45
Hardness (DPN)							
Annealed	45	25	70	25			
Hard	115	58	230	5	4	5	35

the common metals other than silver. Copper is usually classified as a noble metal with high corrosion resistance.

The corrosion reaction generates an electrical potential
between the anodic and the cathodic areas. This electrochemical potential per unit area of interface increases with increasing
ease of dissolution of the metal. Thus metals may be classified
in order of the potential developed. The sequence is sensitive to
the composition of the conducting liquids, and metals close to
each other in the list may interchange places as the acidity or
alkalinity of the liquid is changed. The following list shows the
position of copper and its alloys, relative to other common
metals and alloys, when in contact with seawater:

Anodic
Magnesium and magnesium alloys
Zinc and zinc alloys
Galvanized steel
Aluminium
Duralumin (aluminium, 4.5% copper)
Mild steel
Cast iron
Tin–lead solder
Austenitic stainless steel (active)
Lead
Tin
Muntz metal (60% copper, 40% zinc)
Nickel (active)
Brass (70% copper, 30% zinc)
Copper
Aluminium bronze
Silicon bronze
Monel (67% nickel, 33% copper)
Silver solder (70% silver, 30% copper)
Nickel (passive)
Inconel (76% nickel, 16% chromium, 8% iron)
Austenitic stainless steel (passive)
Silver
Gold
Cathodic

In general, the risk of galvanic corrosion increases with increasing difference in the electromotive potential of the two materials
forming the corrosion couple. *Figure 5.1* illustrates the corrosion compatibility of different pairs of materials where compatibility is defined as a difference of not greater than 0.25 V in the
electrode potentials of the metals or alloys.

The typical mechanical properties at room temperature of
unalloyed copper are shown in *Table 5.3*. Copper in varying
degrees of purity is produced for particular applications.
These range, for example, from (C103) oxygen-free high conductivity copper (99.95% Cu) to (C108) copper cadmium

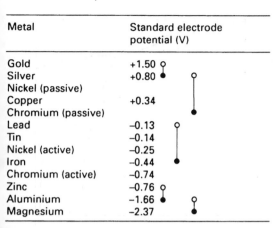

Metal	Standard electrode potential (V)		
Gold	+1.50		
Silver	+0.80		
Nickel (passive)			
Copper	+0.34		
Chromium (passive)			
Lead	−0.13		
Tin	−0.14		
Nickel (active)	−0.25		
Iron	−0.44		
Chromium (active)	−0.74		
Zinc	−0.76		
Aluminium	−1.66		
Magnesium	−2.37		

Figure 5.1 *Compatible metal couples: (○) The most cathodic member of a series; (●) an anodic member*

Table 5.3 *Typical mechanical properties at room temperature of unalloyed copper*

Copper type	0.2% Proof stress $(N\,mm^{-2})$*	Ultimate tensile strength $(N\,mm^{-2})$	Elongation (%)	Hardness (HV)
Annealed	60	220	55	45
Hard	325	385	4	115

*$1\,N\,mm^{-2} \approx 1\,MPa$.

containing 0.7 to 1.3% cadmium. Other coppers contain small proportions of lead, tellurium or sulphur.

5.2 Copper alloys

Copper may be alloyed with a number of other metals such as gold, nickel and zinc. Copper and zinc alloy to form brass in the approximate ratio of 2:1 respectively. *Figure 5.2* gives a guide to the variation of properties in relation to the zinc content. Brass alloys include aluminium brass and manganese brass.

Similarly copper may be alloyed with tin to produce bronze; *Figure 5.3* gives a guide to the variation of properties in relation to the tin content.

Copper may also be allowed with aluminium to produce aluminium bronze; *Figure 5.4* gives a guide to the variation of properties in relation to aluminium content. Bronze may also be produced by alloying copper and beryllium.

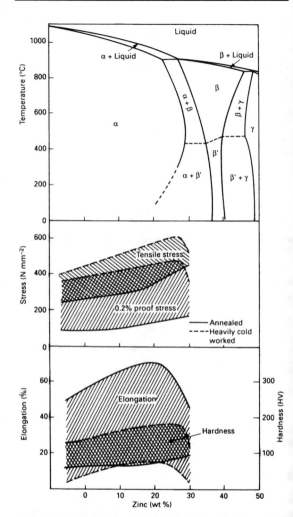

Figure 5.2 *Part of the copper–zinc phase diagram and the variation in mechanical properties with composition for wrought brass*

5.3 Gold and gold alloys

Gold is usually described as the most ductile metal and can be reduced to foil of only $0.1 \mu m$ thickness. The metal is used in construction in electrical and fine finishing applications. Some gold alloy compositions and their colours are illustrated in *Table 5.4*.

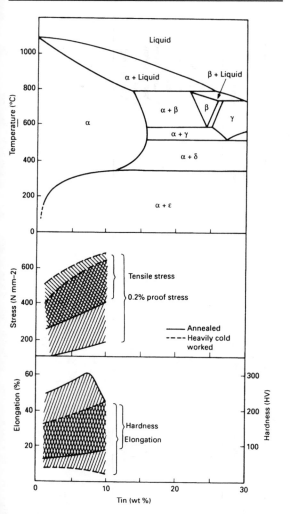

Figure 5.3 *Part of the copper–tin phase diagram and the variation in mechanical properties with composition for wrought tin bronzes*

5.4 Lead and lead alloys

Pure lead is very ductile, has a low melting point, a high coefficient of thermal expansion, poor fatigue strength and a tendency to creep (see *Figure 5.5*). It has a very high density and therefore good damping properties. The creep properties can be varied by alloying the lead with antimony (see *Figure 5.6*).

Table 5.4 Some gold alloy compositions and their colours

Carat	Composition (wt%)					Colour
	Gold	Silver	Copper	Zinc	Nickel	
22	91.7	–	8.3	–	–	Reddish
	91.7	6.2	2.1	–	–	Yellow
	91.7	1.2	7.1	–	–	Deep yellow
18	75.0	9.0	16.0	–	–	Rich yellow
	75.0	20.0	5.0	–	–	Yellow
14	58.3	4.0	31.2	6.4	–	Reddish yellow
	58.3	7.5	26.6	6.6	1.0	Orange yellow
9	37.5	12.5	46.5	3.5	–	Reddish yellow
	37.5	5.5	53.5	3.5	–	Yellowish red

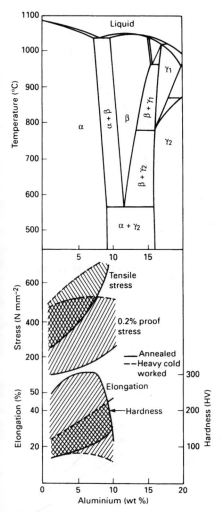

Figure 5.4 *Part of the copper–aluminium phase diagram and the variation in mechanical properties with composition for wrought aluminium bronzes*

Mixtures of lead and bismuth form a series of low-melting point (eutectic type) alloys and the melting point can be depressed further by addition of other elements with low melting points. These alloys are usually called *fusible alloys*; typical compositions and melting points are given in *Table 5.5*.

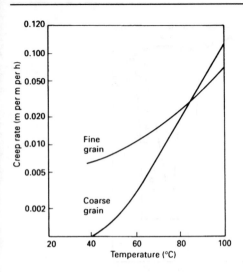

Figure 5.5 *Creep rate of commercial lead under an applied load of 0.49 N mm⁻² in relation to temperature and grain size*

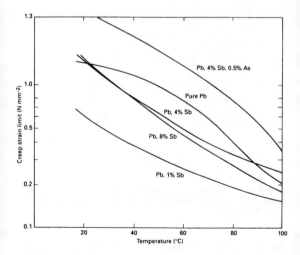

Figure 5.6 *Creep strain limits for wrought pure lead and various lead-antimony alloys producing 1% elongation after 5 × 10⁴ h (5.7 years) under load*

Table 5.5 *Typical compositions and melting points lead-bismuth (fusible) alloys*

Melting point (°C)	Composition (wt%)					
	Pb	Bi	Sn	Cd	In	Sb
46.8	22.6	44.7	8.3	5.3	19.1	–
70	27.3	49.5	13.1	10.1	–	–
						(Quaternary eutectic)
96	32	52.5	15.5	–	–	–
						(Ternary eutectic)
113	25	50	25	–	–	–
247	87	–	–	–	–	13

5.5 Nickel and nickel alloys

Nickel has a white silvery appearance, is a hard, tough malleable metal which exhibits high thermal and electrical conductivities. Its mechanical properties are similar to those of low carbon steel and it is ferro-magnetic. It is commonly used in alloys with copper of which perhaps the best known is the Monel group. Typical mechanical properties are shown in *Table 5.6*.

Table 5.6 *Typical mechanical properties of nickel alloys*

	0.2% Proof stress ($N\,mm^{-2}$)	UTS ($N\,mm^{-2}$)	Elongation (%)	Hardness (VDN)
Monel 400 annealed	250	520	45	140
Monel K 500 annealed	500	650	35	270
Annealed and aged	700	1000	20	300

5.6 Silver and silver alloys

Silver has the highest reflectivity of all metals, reflecting about 95% of all but ultraviolet light. Although it has lower resistance to oxygen than gold, it does not normally oxidize at ambient temperatures. For decorative finishes the coating thickness is usually in the range of 10–40 μm. It is also used in heating coils, stills and condensers in the pharmaceutical industry where the absence of metallic contamination is critical.

5.7 Tin

Tin is usually described as a soft metal. The pure metal is white with a bluish tinge and high lustre which if heated in air to above 170°C forms a thin layer of hard yellow oxide which prevents further oxidation. In construction it is mainly used as a hot-dip coating of steel and as solder (see *Figure 5.7*).

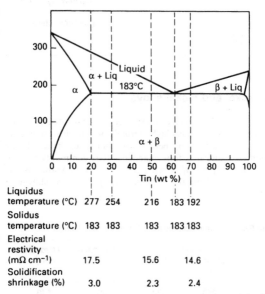

Liquidus temperature (°C)	277	254	216	183	192
Solidus temperature (°C)	183	183	183	183	183
Electrical restivity (mΩ cm⁻¹)		17.5	15.6	14.6	
Solidification shrinkage (%)		3.0	2.3	2.4	

Figure 5.7 *The tin–lead phase diagram and some properties of common solders*

5.8 Zinc

In terms of tonnage used, Zinc is the fourth most important metal (behind iron, aluminium and copper). Half the production is used for coating other metals, much of the remainder being used for alloying with aluminium, copper and magnesium. The melting point of zinc is higher than that of lead and tin but it boils at 911°C. Since zinc is electronegative to the common engineering materials (such as steel) it provides true galvanic protection to these metals and does not act solely as a corrosion barrier as is the case for example with tin and lead coatings on steel. Typical lifetimes of hot-dip zinc coatings are shown in *Table 5.7*. The most common form of zinc coating is hot-dip galvanizing of unalloyed low carbon steel. Zinc is also produced as wrought sheet for roofing and

Table 5.7 *Typical lifetimes (years) in relation to thickness of hot-dip zinc coatings*

Coating thickness (μm)	Atmosphere		
	Rural	Marine	Severe industrial
5	3	1	0.5
25	14	7	3
50	30	13	6

*Zinc coatings are often specified by weight: $100 \, g \, mm^{-2} \equiv 0.3 \, oz \, ft^{-2} \equiv 14.1 \, \mu m$.

cladding. A relatively recent development is the production for these purposes of zinc sheet containing at least 0.05% titanium to increase creep resistance and at least 0.10% copper to improve the mechanical strength. Typical properties are shown in *Table 5.8*.

Table 5.8 *Typical properties of zinc-copper titanium sheet*

	Rolling direction	Transverse direction
Minimum UTS ($N \, mm^{-2}$)	150	200
Minimum elongation (%)	30	15

Bibliography

BODSWORTH, C.R.B. In DORAN, D.K. (ed.), *Construction Materials Reference Book*, Butterworth-Heinemann, Oxford, Chapter 7 (1992)

Part Two: Non-metals

6 *Asbestos*

6.1 Introduction

Since the first recorded use of asbestos, the mineral has become progressively more popular as its many physical and chemical properties became known. The range of products in common use which contain asbestos is very large. The common perception of asbestos as the corrugated sheeting on garage and factory roofs is merely the tip of the iceberg. The applications of asbestos in the construction industry are no less diverse.

Asbestos is the general term used to describe a number of naturally occurring crystalline metal silicates which, because of the array of physical and chemical properties they possess, such as their resistance to chemical attack, immense thermal stability and good tensile strength, have become extremely useful materials for use in many industrial and construction applications. The minerals, when crystallized, form narrow parallel rods in tight groups or bundles. These minerals, if physically stressed, will break not only into shorter lengths, but will split into progressively finer fibrils. It is this particular facet which is the major reason for the mineral's array of physical properties.

There are six main types of asbestos, namely chrysotile (commonly known as white asbestos), amosite (known as brown asbestos), crocidolite (blue asbestos), anthophyllite, actinolite and tremolite. Several other minerals are classed as asbestos but have not been commercially used for various reasons. Asbestos can be classified into two groups, the serpentines and the amphiboles, labels which refer primarily to aspects of their crystalline structure (see *Table 6.1*).

Chrysotile is the only popularly used serpentine mineral, though other non-fibrous asbestos types, lizardite and antigorite, are also serpentines. The serpentine group is distinct from the amphibole group in that their crystalline structures are formed in layer lattices. Chrysotile appears white to the naked eye, hence its popular name of white asbestos.

Amphiboles, which include the remainder of the asbestos types mentioned, are chain silicates which means that, in contrast to the layered crystalline formation of the serpentines, they form preferentially along the lengthwise direction of the fibres, giving them a strong lengthwise tensile strength. Crocidolite and amosite are the only amphibole minerals to have been mined in significant quantities. Visually, crocidolite appears deep blue to turquoise in colour and amosite a distinctive grey brown.

Table 6.1 *Asbestos types, chemical formulae and physical properties*

Asbestos type	Approx. chemical formula	Asbestos group	Colour	Physical properties
Chrysotile	$3MgO, 2SiO_2, 2H_2O$	Serpentine	White/pale green	Alkaline resistance: good Acid resistance: poor Heat resistant to: 650°C Average density: 2.5×10^3 kg m^{-3} Tensile strength: very good
Amosite	$5.5FeO, 1.5MgO, 8SiO_2, H_2O$	Amphibole	Pale brown	Alkaline resistance: good Acid resistance: fair Heat resistant to: 800°C Average density: 3.3×10^3 kg m^{-3} Tensile strength: good
Crocidolite	$Na_2O, Fe_2O_3, 3FeO, 8SiO_2, H_2$	Amphibole	Blue/turquoise	Alkaline resistance: good Acid resistance: good Heat resistant to: 650°C Average density: 3.3×10^3 kg m^{-3} Tensile strength: excellent
Anthophyllite	$7MgO, 8SiO_2, H_2O$	Amphibole	White/pale brown	Alkaline resistance: good Acid resistance: excellent Heat resistant to: 800°C

			Properties
Actinolite	$2CaO, 4MgO, FeO, 8SiO_2, H_2O$	Amphibole	Pale green	Average density: 3.0×10^3 kg m^{-3} Tensile strength: poor Alkaline resistance: fair Acid resistance: fair Heat resistant to: 900°C
Tremolite	$2CaO, 5MgO, 8SiO, H_2O$	Amphibole	White	Average density: 3.1×10^3 kg m^{-3} Tensile strength: poor Alkaline resistance: good Acid resistance: excellent Heat resistant to: 1000°C Average density: 3.0×10^3 kg m^{-3} Tensile strength: poor

Chrysotile, because of its crystalline structure, has a very soft nature when broken in fibre tufts, thus making the material ideal for weaving. Amphiboles, and in particular amosite, are very rigid and hence unsuitable for weaving, but have excellent thermal properties which have been utilized in a variety of insulation materials.

In the 1970s the extent to which even casual exposure to asbestos could cause health damage was realized. By 1985 its use in the West had been severely curtailed with a virtual ban on the importation and use of crocidolite and amosite. *Figure 6.1* clearly indicates production trends. Current production is around 5 million tonne year^{-1}. However, with the increasing worldwide ban on the use of asbestos future production seems set for a further fall, although its use in the third world will persist for the foreseeable future.

6.2 Application to construction

Extensive use of asbestos occurred after 1910 with the introduction of asbestos cement sheeting. Chrysotile rope was also used for gaskets on boilers and in other forms of insulation of hot water services.

Chrysotile was also sorted into long fibres and woven into fabrics for use as fire blankets and, in its finer form, for cable insulation. These products, when abraded, discard large numbers of asbestos fibres into the atmosphere thus causing a serious health hazard.

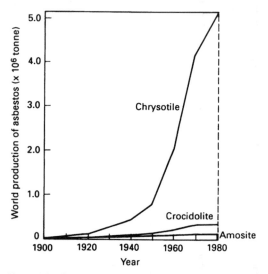

Figure 6.1 *Growth in world production of asbestos to 1980 with production levels for chrysotile, amosite and crocidolite*

A guide to the various types of products and their likely asbestos content is shown in *Table 6.2* and the appropriate British Standards are shown in *Table 6.3*.

6.3 Dealing with existing asbestos

As mentioned earlier, human contact with asbestos fibres constitutes a serious health hazard. Regulation of the use of the material in the UK has become increasingly prohibitive. These restrictions began with the Asbestos Regulations (1969) and have progressed to the point where the new use of the material is virtually banned. However, a great deal of asbestos is present in existing construction and precautions for dealing with this are necessary. The current approach is as follows:

- Removal under strictly controlled conditions, or
- Encapsulation to avoid detachment of fibres.

A summary of current legislation is indicated in *Table 6.4*.
Successive legislation has reduced the level of fibres in the atmosphere thought to be safe for workers without respiratory protection. In 1984 these levels were stated as:

- Chrysotile 0.5 fibre ml^{-1}
- Crocidolite and amosite 0.2 fibre ml^{-1}

In general, however, all workers should be shielded from the effects of asbestos fibres and should contact be necessary even at the lowest levels of exposure then the wearing of respiratory protection is strongly recommended.

Table 6.2 *Typical asbestos content and types used in common asbestos products and materials*

Asbestos material	Probable asbestos content (%)	Main asbestos types used*	Other asbestos types used*
Textured paint	2–12	Ch	A
Bituminous felt	3–20	Ch	
Thermoplastics	5–20	Ch	A
Asbestos cement	10–20	Ch	C
Insulation board	15–40	A	Ch, C
Asbestos millboard	30–65	A	Ch
Gasket material	30–60	Ch	An, A
Brake linings	30–60	Ch	
Thermal insulation	20–75	Ch, C, A	
Sectional insulation	30–80	Ch, A	C
Spray coatings	60–90	A, C	Ch
Asbestos textiles	60–100	Ch	C

*Ch, chrysotile; C, crocidolite; A, amosite; An, anthophyllite.

Table 6.3 *British Standards and associated documentation for asbestos products*

Product	Standard
Asbestos cement	
Boards impregnated for electrical purposes	BS 3497: 1979 (1986)
Building products—test methods	BS 4624: 1981
Corrugated sheet roof and wall covering	BS 5247: Part 14 (1975)
Flue pipes and fittings (light quality)	BS 567: 1973 (1989)
Flue pipes and fittings (heavy quality)	BS 835: 1973 (1989)
Pipes and fittings for sewerage and drainage	BS 3656: 1981
Pressure pipes and joints	BS 486: 1981
Rainwater goods	BS 569: 1973 (1987)
Slates and sheets for building	BS 690: Parts 1–4 (1973–1981)
Cisterns	BS 2177: 1973 (withdrawn)
Asbestos cement decking	BS 3717 (1972) (withdrawn)
Soil, waste and ventilating pipes and fittings	BS 582: 1965
Asbestos cement pipelines	
Field pressure testing	BS 5886: 1980
Guide for laying	BS 5927: 1980
In land	CP 2010: Part 4, 1972
Asbestos insulating board	
Asbestos insulating boards and asbestos wallboards	BS 3536 (withdrawn)
Sprayed asbestos	
Sprayed asbestos insulation	BS 3590 (withdrawn)
Others	
PVC (vinyl) asbestos floor tiles	BS 3260: 1969
Bitumen sheet roof coverings	CP 143 (16)
Asbestos packed sheet conduit	BS 731: Part 1, 1952 (1980)
Untreated asbestos paper (for electrical purposes)	BS 3057: 1958
Woven tape for electrical insulation	BS 1944: 1973 (1984)
Compressed asbestos fibre jointing	BS 2815: 1973
Compressed asbestos fibre jointing (oil resistant)	BS 1832: 1972
Compressed asbestos fibre jointing (rubber bonded for aircraft)	Aero FD5

The removal of asbestos should only be carried out by contractors who are licensed to carry out such work. It is a costly operation often involving not only the use of protective clothing but also the provision of sealed and air-locked working enclosures. The aim should be to remove all asbestos; in practice a level not greater than 0.01 fibre ml^{-1} is acceptable.

Encapsulation techniques vary with the type of asbestos being considered. Typical examples include:

- Board products—by covering externally with bitumen-backed canvas and internally with a paint seal.
- Pipe insulation—shrouding with plastic, stainless steel, aluminium hammer cladding, calico and canvas covering or solid plaster casing.
- Sprayed coatings—low-pressure spraying with thick, fire-retardant elastomeric sealant. High-pressure spraying may dislodge or damage the existing coatings.

Table 6.4 *Summary of current legislation, Codes of Practice and Guidance Notes for work with asbestos*

Legislation
The Control of Asbestos at Work Regulations (1987)
The Asbestos (Licensing) Regulations (1983)
The Asbestos (Prohibition) Regulations (1985)
The Health and Safety at Work, etc., Act (1974)

Codes of Practice
Approved Code of Practice—The Control of Asbestos at Work
Approved Code of Practice—Work with Asbestos Insulation, Asbestos
                                    Coating and Asbestos Insulating Board

Guidance notes—Environmental Hygiene Series
EH 10	Asbestos: Exposure Limits and Measurement of Airborne Dust Concentrations
EH 35	Asbestos Dust Concentrations at Construction Processes
EH 36	Work with Asbestos Cement
EH 37	Work with Asbestos Insulating Board
EH 41	Respiratory Protective Equipment for Use Against Asbestos.
EH 47	The Provision, Use and Maintenance of Hygiene Facilities for Work with Asbestos Insulation and Coatings

6.4 Health problems

It is beyond the scope of this book to deal in any depth with illnesses that may be caused by exposure to asbestos fibres. The following are, however, some of the relevant complaints, many of which affect the lungs:

• fibrosis of the lungs—usually known as asbestosis
• lung cancer
• mesothelioma—a highly malignant tumour
• skin complaints
• other cancers.

For anyone dealing with asbestos it is imperative to seek expert advice.

Bibliography

HARRIS, R. In DORAN, D.K. (ed.), *Construction Materials Reference Book*, Butterworth-Heinemann, Oxford, Chapter 9 (1992)

7 Bituminous materials

Bitumen is a mixture of hydrocarbons which are soluble in carbon disulphide. It occurs naturally or can be distilled from petroleum. The material has a high density and waterproofing qualities. It is used in asphalt for tanking basements and waterproofing roofs. For further information on these applications the reader is referred to publications by BACMI and NAM & MC. Bitumen is also a constituent of roofing felt to BSCP144 Pt 3. This chapter, however, deals with asphalts and coated macadams used in road construction.

In the USA and many other parts of the world, any bituminous material is described as an asphalt. In the UK, asphalts have specifically been gap-graded materials with relatively high bitumen contents. They derive a significant part of their strength from the bitumen and are produced for use in wearing courses, base courses or road bases. The combination of high bitumen content and the stiff bitumen used in their manufacture results in their generally only being laid successfully at relatively high temperatures by paving machines. The relevant British Standard, BS 594: 1985, has two parts, the first covering the specification for constituent materials and asphalt mixtures and the second dealing with the transport, laying and compaction of rolled asphalts.

The most commonly used type of rolled asphalt wearing course typically contains approximately 55% sand, 30% coarse aggregate and 15% bitumen binder.

In 1985, a revised edition of BS 594 was published and this contains four categories of wearing-course mix:

(1) Those produced using primarily natural sand fines with binder contents determined from Marshall tests for particular ranges of stability, and designated 'design type F'
(2) Other design mixes based on crushed-rock fines, of coarse sands called 'design type C'
(3) 'Type F' recipe mixes
(4) 'Type C' recipe mixes.

The last two categories are virtually the same recipe mixes used since the early 1970s when crushed-rock fines were first permitted, and the type F recipe mixes are little changed from those used since before World War I, except that they have lower binder contents.

Two of the tables from BS 594 are reproduced here (*Tables 7.1* and *7.2*) to indicate the composition of type F and type C wearing-course mixtures. The Marshall test records the compressive strength or 'Marshall stability' of scientifically prepared samples. Typical Marshall test results are shown in

Table 7.1 Partly from BS 594: Part 1: 1986: Section 3: Table 3. Composition of design type F wearing-course mixtures

Column number	7	8	9	10	11	12
Designation*	0/3†	30/10	30/14	40/14	40/20	55/20
Nominal thickness of layer (mm)	25	35	40	50	50	50
Percentage by mass of total aggregate passing BS test sieve						
28 mm	—	—	—	—	—	100
20 mm	—	—	—	—	100	90–100
14 mm	—	—	100	100	95–100	35–80
10 mm	—	100	85–100	90–100	50–85	—
6.3 mm	100	85–100	60–90	—	—	—
2.36 mm	95–100	60–72	60–72	50–62	50–62	35–47
600 µm	80–100	45–72	45–72	35–62	35–62	25–47
212 µm	25–70	15–50	15–50	10–40	10–40	5–30
75 µm	13.0–17.0	8.0–12.0	8.0–12.0	6.0–10.0	6.0–10.0	4.0–8.0
Maximum percentage of aggregate passing 2.36 mm and retained on 600 µm BS test sieves	22	14	14	12	12	9
Minimum target binder content % by mass of total mixture‡	9.0	7.0	6.5	6.3	6.3	5.3

*The mixture designation numbers (e.g. 0/3 in column 7) refer to the nominal coarse aggregate content of the mixture/nominal size of the aggregate in the mixture, respectively.
†Suitable for regulating course.
‡In areas of the country where prevailing conditions are characteristically colder and wetter than the national average the addition of a further 0.5% of binder may be beneficial to the durability of the wearing courses.

Table 7.2 *Partly from BS 594: Part 1: 1985: Section 3: Table 4. Composition of design type C wearing-course mixtures*

Column number	13	14	15	16	17
Designation*	0/3	30/10	30/40	40/14	40/20
Nominal thickness of layer (mm)	25	35	40	50	50
Percentage by mass of total aggregate passing BS test sieve					
28 mm	–	–	–	–	100
20 mm	–	–	–	100	95–100
14 mm	–	100	100	90–100	50–85
10 mm	–	85–100	85–100	50–85	–
6.3 mm	100	60–90	60–90	–	–
2.36 mm	90–100	60–72	60–72	50–62	50–62
600 µm	30–65	25–45	25–45	20–40	20–40
212 µm	15–40	15–30	15–30	10–25	10–25
75 µm	13.0–17.0	8.0–12.0	8.0–12.0	6.0–10.0	6.0–10.0
Minimum target binder content % by mass of total mixture†	9.0	7.0	6.5	6.3	6.3

*The mixture designation numbers (e.g. 0/3 in column 13) refer to the nominal coarse aggregate content of the mixture/nominal size of the aggregate in the mixture, respectively.
†In areas of the country where prevailing conditions are characteristically colder and wetter than the national average the addition of a further 0.5% of binder may be beneficial to the durability of the wearing courses.

Figure 7.1 and the required Marshall stability versus traffic flow (commercial vehicles per lane per day) in *Table 7.3*.

7.1 Asphalt base courses

Hot-rolled asphalt base courses are always manufactured as recipe mixes using 50 pen grade bitumen and a variety of gap-graded natural aggregates. Although the binder content of a typical base course is significantly lower than that of the wearing course with which it will be overlaid, it is nevertheless almost 50% greater than in a dense macadam base course. This, coupled with the much higher fines content of the asphalt, makes it a very dense, impermeable mixture with high fatigue strength and the combination of both hot-rolled asphalt base course and wearing course considerably enhances the strength

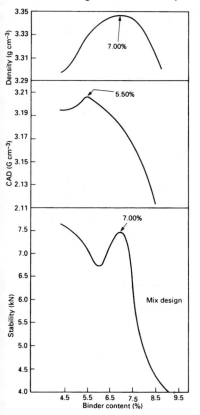

Figure 7.1 *Typical Marshall test results. (CAD = Compacted aggregate density)*

Table 7.3 *Partly from BS 594: Part 1: 1986: Appendix B.2: Table 11. Criteria for the stability of laboratory design asphalt*

Traffic (cvpd)	Marshall stability of complete mix (kN)
<1500	2 to 8*
1500–6000	4 to 8
>6000	6 to 10

*It may be necessary to restrict the upper limit where difficulties in the compaction of materials might occur.
Note 1. For stabilities up to 8.0 kN the maximum flow value should be 5 mm. For stabilities in excess of 8.0 kN a maximum flow of 7 mm is permissible.
Note 2. The stability values referred to should be obtained on laboratory mixes.

of any road pavement. This is clearly illustrated in *Figure 7.2* which shows the superior performance of these materials relative to the admittedly very much weaker macadams then in use, which were manufactured with softer bitumens than is now normal, i.e. 300 pen.

7.2 Asphalt road and airfield bases

Current DTp practice is to require the same thickness of roadbase construction in flexible pavements, whether built using hot-rolled asphalt road base or dense bitumen macadam road base. This is a relatively recent practice and, until 1985, the higher binder content of the rolled asphalt, coupled with its stiffer bitumen and higher fines content, was always acknowledged as producing the stiffer of the two materials. Hence, although the rolled-asphalt road base was more expensive, a lesser thickness was required. For example, the most recent Road Note 29, valid until 1985, required a 120 mm thickness of dense bitumen macadam but only a 100 mm thickness of rolled-asphalt roadbase to carry 4 million standard axles.

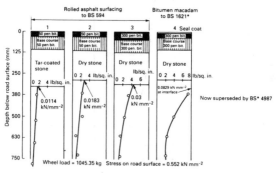

Figure 7.2 *Effect of type of road structure on vertical stress on subgrade. (Tests by TRL)*

.3 Asphalt regulating courses

Whilst the laid thickness of bituminous pavement layers can be varied within limits as they are laid, it is often necessary to reshape them either as part of remedial works or in marrying new works to existing road pavements. The reshaping or 'regulating' is carried out using either a 'sand carpet' or thin base-course material. Sand carpet is merely hot-rolled asphalt wearing course without its course aggregate, and this can be laid down to a feather-edge. For thicknesses in excess of 10 mm it is usual to use a base-course asphalt with a maximum aggregate size of 10 mm. Older bituminous surfaces need to be 'tack-coated' using a cold-applied K-1-40 bituminous emulsion to assist the bonding of the regulating course, which may then be overlaid as soon as it has cooled to ambient temperature.

.4 Medium-temperature asphalts for use in wearing courses

Medium-temperature asphalts (MTAs) are those materials produced using 100 pen grade bitumens. These were originally produced by major bituminous-material suppliers under various trade names as cheaper alternatives to hot-rolled asphalt, but with substantially greater durability than bituminous macadams. The materials were laid in relatively thin layers 25–30 mm) with a nominal application of chippings on housing estates in many large cities, carrying very few, if any, vehicles larger than a double-decker bus, and gave extremely good service with lives of up to 20 years before they needed any maintenance. Their use was extended by some county highway authorities onto lightly trafficked rural roads, with stone contents up to 55%, and with no precoated chippings. With a maximum aggregate size of 20 mm these had about the same texture as the lightly chipped housing-estate roads and provided good skid resistance at traffic speeds of less than 50 miles h^{-1}.

.5 Mastic asphalt in wearing courses

Mastic asphalt is an extremely expensive material by virtue of its very high binder content. It is now rarely, if ever, used in carriageways or footways, except in high-prestige areas and in reinstating existing surfaces which tend to be extremely durable due to the lack of oxidation of the binders. Mastic asphalt is produced by mixing approximately equal volumes of coarse aggregate, filler and hard bitumen in stirred, heated tanks, from which the material is removed as a molten liquid, in buckets. It is then discharged onto the surface to be overlaid and the material spread using wooden floats. Footway surfaces are finished with a crimping pattern roller. Carriageways are finished with an application of hand-spread precoated chippings which are lightly rolled-in.

The relevant standards are BS 1446 for natural rock aspha
fine aggregate, and BS 1447 for mastic asphalt produced usin
limestone aggregate.

7.6 Coated macadams

The concept of stone-to-stone contact and interlock which
fundamental to all macadams, is attributed to McAdam,
Scottish civil engineer. He is understood to have adopted th
principle in rebuilding existing roads which had been forme
from much larger pieces of stone, but nevertheless failed.

Today coated macadams are perceived as having the funda
mental aggregate interlock, which gives them their resistance t
deformation and load-spreading ability, but also some tensi
strength derived from the binder which gives them resistance t
fatigue and, for the denser macadams at least, a degree o
impermeability to water which assists in protecting the sub
grade soils. One of the prime functions of the binder is to ac
as a lubricant at laying stage, and assist in breaking down th
intergranular friction, which in turn further enhances the aggre
gate interlock.

BS 4987 contains the grading and binder requirements for 1
coated macadam mixtures including two roadbase material
five base-course mixtures and nine wearing courses.

7.7 Roadbase macadams

The load-carrying capacity of any bituminous road pavemer
and its ability to relieve the natural subgrade of critical value
of vertical strain depends primarily on the stiffness or elasti
modulus of the roadbase. A roadbase material which deform
in the wheel-tracked zones has too low a modulus; however, th
in situ performance of the material depends not only on th
written specification but also on the manufacturer producin
it to specification, and its being transported, laid and com
pacted on a stable construction platform, sufficient to achiev
95% of the percentage refusal density (PRD). Assuming a
these requirements are achieved, the engineer may elect to us
either 40 mm dense bitumen macadam (DBM) in his road bas
or DBM with a maximum aggregate size of 28 mm.

7.8 Base-course macadams

There are now six base-course macadams: one open graded an
three dense graded specified in BS 4987: 1988 and the tw
improved base-course macadams originally specified in th
DTp specification, but now included in BS 4987 by amendmen

7.9 Wearing-course macadams

BS 4987: 1988 contains specifications for two open-grade
wearing-course materials, and one medium-graded, one fine

graded, one dense-graded, two close-graded and two pervious macadams.

Appendix B to BS 4987: 1988 stresses the need to ensure that the coarse aggregates in any macadam wearing course used on carriageways are resistant to polishing, particularly on high-speed roads and at junctions and on roundabouts. This advice is particularly relevant in view of the DTp's recent emphasis on providing better skidding resistance on trunk roads, and the DTp publications HA/36/87 and HD/15/87. It also draws attention to the rate at which aggregates are abraded by heavy traffic in particular and refers to guidance published in DTp Technical Memorandum H16/76. Only fine-grade wearing courses are likely to be finished with an application of precoated chippings.

7.10 Recycled materials

The oil crisis of the early 1970s and consequent rapid escalation in the cost of bitumen focused attention on the possibility of recycling bituminous materials, usually cold planings removed from road pavements in maintenance operations. A hot recycling operation must be used for asphalts and dense macadams, but cold recycling can be used for lower grade materials being either planed asphalts or macadams.

In the hot process, the recycled aggregates, which already contain a substantial mass of bitumen suitable for re-use, cannot be heated by passing them through a direct flame-heated high-temperature zone into the drier since this would simply burn and destroy the binder. Instead, therefore, they are fed into a 'drum mixer' beyond the burner, and heated by direct contact with the virgin aggregate which has already been flame heated. Therefore, the process is as for virgin materials except that less new bitumen is added to the mix, which makes it more economical than material produced without recycled materials. Where the binder on the recycled aggregates is unsuitable for re-use, these may simply be added to virgin aggregate in a batch mixer and then coated with new binder. The DTp and the Department of Energy are concerned to encourage the use of recycled materials as this not only reduces the consumption of valuable natural resources (bitumen and aggregate), but also requires less energy in the total production process. The DTp is, however, insistent in requiring that mixtures which incorporate recycled materials achieve the same specification compliance as products made with 100% virgin raw materials.

An *in-situ* process known as 'Repave' is described in DTp document HA/14/82.

Bibliography

FARRINGTON, J.J. In DORAN, D.K. (ed.), *Construction Materials Reference Book*, Butterworth-Heinemann, Oxford, Chapter 10 (1992)

8 Ceramics

8.1 Bricks and brickwork

The term 'ceramics' is used to describe the plastic arts of the potter and other workers in clay, and is derived from the Greek 'keramos' meaning 'potter's clay'. It is also interesting to note that the Latin word for tile is 'tegula' from the verb 'tego' meaning to cover. The art of producing ceramics from indigenous materials, especially earthenware, has roots in ancient civilizations sited around the Mediterranean, as demonstrated by archaeological discoveries of pottery and other clay products used as decoration, utensils and building materials.

Bricks are probably the oldest industrialized building material known to man. Surviving examples of brickwork from *ca.* 1300 BC still maintain an attractive appearance. The earliest bricks were made from clay, taken from close to the surface of the ground, or from river banks, moulded into shape by hand and dried in the sun. 'Adobes' as they are known have been found in the remains of Jericho dating back to about 8000 BC.

Clay bricks over the centuries were traditionally made locally and not transported very far, so that they had widely differing characteristics depending on the material available and the way in which the bricks were treated by the maker; it was quite usual for the bricks to be made on the building site, so that they did not need to be transported at all—assuming that there were suitable supplies of soft mouldable clay available. Much of the attraction of brickwork lies in the textures, colours and variations that arise from the use of materials made from various clays and manufacturing processes.

In the 19th century a process was developed for making bricks from sand and lime by subjecting them to high-pressure steam (autoclaving) to bind the materials together by the formation of calcium silicate. Additionally, bricks are now made from concrete, and there are even plastic imitations. Other materials may be added to clay in the manufacturing process to improve the properties, for instance sand, chalk or chemical admixtures.

In Britain, a brick has traditionally been of such a size that it can be picked up in one hand, to be laid on a bed of mortar. Modern processes permit clay to be formed into much larger units and, whilst this is done in other countries, it is not practised in the UK. Again in other countries, large calcium silicate units are manufactured and used but they are not made in the UK.

In addition to being used in walls, either for their attractive facing appearance, or as part of the structural support system, bricks are used as features in landscaping work and as hard paving, both inside and outside buildings.

Calcium silicate brick production requires a supply of suitable sand and this is not so readily available in the UK as it is in some continental countries. Some calcium silicate bricks use a coarse aggregate as well as sand, and these are usually known as flint-lime bricks, as opposed to sand-lime. Apart from sand, aggregate and lime, other admixtures may be used in the process, particularly colouring agents; the final process is autoclaving.

8.1.1 Types of brick

Bricks are classified both by their quality as it would affect their use, and the manufacturing process as it affects their appearance. BS 6100: Section 5.3: 1984: Glossary of Terms gives extensive definitions of the relevant words, but a distinction needs to be drawn between the following usage categories.

Common bricks   Common bricks are those that are used where they will not normally be exposed to view and where there is no claim as to their appearance. They are suitable for general use in construction, given that they possess adequate strength and durability for the location. The term relates to appearance only and has no significance in the classification of bricks in BS 3921: 1985.

Facing bricks   Facing bricks have a suitable appearance for use where they will be exposed to view so that they give an attractive and pleasing effect. Very many types of facing brick are available and the choice can be made either from catalogues and display panels or from examples of work carried out in the particular brick.

Common terms to describe appearance are: sandfaced; rustic; multi-coloured; blue/yellow/red etc, and smooth. The designation of a brick as facing does not imply that it will be durable in all locations.

Engineering bricks   The term 'engineering bricks' is used almost only in Britain. These bricks may be facing bricks, but are not defined with facing qualities in mind. Engineering bricks are required to conform to defined upper limits for water absorption and lower limits for compressive strength as specified in BS 3921: 1985. The bricks are used where high strength and/or resistance to aggressive environments is required.

Damp-proof-course bricks   Damp-proof bricks are ones with defined upper limits to water absorption, similar to engineering bricks, but without the same strength requirements. Two courses of such bricks are permitted to fulfil the function of a damp-proof course according to BS 743: 1970 and are particularly useful at the base of free-standing and retaining walls, where the continuity of the brickwork is maintained without the plane of weakness that a sheet damp-proof course introduces.

8.1.2 Terms that describe bricks by the manufacturing process

These include: extruded wire cut; pressed; soft-mud; stock; hand-made and specials. Forms in regular production are described in BS 4729: 1971.

8.1.3 Properties of bricks

Size   The limits on size for clay and calcium silicate bricks are shown in *Tables 8.1* and *8.2* respectively. For comparison the permitted deviations for concrete bricks are shown in *Table 8.3*.

Water absorption   The water absorption of a clay brick is the percentage of the increase in weight of a dry brick when it has been saturated. It is one of the parameters for the definition of engineering bricks and damp-proof-course bricks. The water absorption of the bricks used in a wall affects the mode of rain penetration through the outer leaf of a cavity wall and is used to define the flexural strength used in lateral load design.

Table 8.1   *Limits of size for clay bricks*

Specified dimension (mm)	Overall measurement of 24 bricks (mm)	
	Max.	Min.
65	1605	1515
102.5	2505	2415
215	5235	5085

Table 8.2   *Limits on manufacturing size of calcium silicate bricks*

	Length	Width	Height
Max. limit of size (mm)	217	105	67
Min. limit of size (mm)	212	101	63

Table 8.3   *Dimensional deviations of concrete bricks*

	Maximum dimensional deviation (mm)	
Length	+4	−2
Height	+2	−2
Thickness	+2	−2

Compressive strength The compressive strength of a brick is the mean of 10 crushing tests, when the failing load is divided by the gross area of the brick. Different types of clay bricks have compressive strengths ranging from about 7 to well over $100\,\text{N}\,\text{mm}^{-2}$ and calcium silicate and concrete from about 21 to nearly $60\,\text{N}\,\text{mm}^{-2}$.

Soluble-salt content Most clays used in brick-making contain soluble salts that may be retained in the fired bricks. If brickwork becomes saturated for long periods, soluble sulphates may be released. These may cause mortars that have been incorrectly specified or batched and have a low cement content to deteriorate under sulphate attack. Of course sulphates from the ground or other sources may be equally destructive. Some clay bricks meet limits placed on the level of certain soluble salts in BS 3921: 1985 and these are designated 'L', signifying low soluble-salt content. Those that do not meet the limits are designated 'N' for normal soluble-salt content.

Frost resistance The British Standard for clay bricks (BS 3921: 1985) contains no mandatory test method for classifying frost resistance, and manufacturers are required to state the frost resistance of their clay bricks by classifying them from experience in use as:

F which means frost resistant, even when used in exposed positions where the bricks will be liable to freezing while saturated.
M which means moderately frost resistant and suitable for general use in walling that is protected from saturation.
O which means not frost resistant.

The frost resistance of calcium silicate bricks is generally higher for the higher strength classes. However, calcium silicate bricks should not be used where they may be subject to salt spray, e.g. on sea fronts, or to de-icing salts.

Efflorescence Efflorescence is a crystalline deposit left on the surface of clay brickwork after the evaporation of water carrying dissolved soluble salts. Manufacturers have to state the category to which the bricks being offered correspond when subjected to the efflorescence test described in BS 3921: 1985, namely the categories nil, slight or moderate. Specifiers should be aware that these categories do not relate to the degree of efflorescence to which brickwork may be liable under certain site conditions. The risk of efflorescence, which is harmless and usually temporary, can best be minimized by protecting both the bricks in the stacks as well as newly built brickwork from rain (see BS 5628: Part 3: Clause 3.5: 1985).

Thermal movement Bricks expand on being warmed, and shrink when cooled. In practice it is the expansion and contraction of the brickwork, rather than of the individual units, that is of interest to the user. Some reversible movement will occur in brickwork exposed to temperature variation. The type of brick will not make any significant difference and the coefficient of thermal expansion in a horizontal direction will be about 5.6×10^{-6} for a 1°C temperature change. For a 10 m length of wall, variation in length between summer and winter might be about 2 mm. Vertical expansion may be up to 1.5 times the horizontal value. *BRE Digest 228* lists thermal movements for many materials.

Moisture movement Clay bricks expand on cooling from the kiln, as some of the molecules of water reattach themselves after being driven off by the heat of the kiln. This expansion is effectively non-reversible, unless the bricks are refired. The magnitude of this movement varies according to the type of brick. Fortunately, a considerable portion of the expansion takes place quite quickly, probably at least half of it occurring within a few days. The remainder may take place slowly over a considerable period (see *BRE Digest No. 228*).

In addition, there is a small reversible movement due to wetting and drying of clay bricks. Calcium silicate bricks tend to shrink as they dry out after manufacture, and then expand again if wetted. The drying shrinkage may range from less than 0.01% to about 0.04%. Limits for drying shrinkage are no longer specified in BS 187: 1978.

Concrete bricks behave like other concrete products, e.g. blocks, in that they shrink on drying and expand on wetting. BS 6073: Part 1: 1981 gives a limit on the drying shrinkage that can be permitted of 0.04%.

As with thermal movement, it is the movement of the brickwork itself that is of interest to the user. Moisture movement in bricks may occur in the form of an irreversible movement which continues throughout the life of the building. Very occasionally, there can be a third type of movement, occurring as a continuing expansion in which bricks expand on drying and expand again if wetted. The amount of information on this type of movement is limited but it is clear that bricks should not be delivered straight from the kilns and used immediately.

8.1.4 Mortar for brickwork
Guidance is given in *Table 8.4*.

8.1.5 Properties of brickwork
Brickwork has extremely good load-bearing qualities and many successful high-rise structures have been built relying only on the brickwork to carry the load to the ground. The detailed design of such walls is beyond the scope of this book, but BS 5628: Part 1: 1978 gives detailed information, and is supplemented by the *Handbook to BS 5628: Part 1*, published by BDA.

Brickwork is also used in conjunction with reinforcement or prestressing tendons to act as a strong and economic

Table 8.4 Mortar mixes

Mortar designation	Type of mortar† (proportions by volume)	Air-entrained mixes‖	
	Cement: lime: sand†§	Masonry cement: sand‡	Cement: sand with plasticizer‡
(i)	1:0 to ¼:3		
(ii)	1:¼:4 to 4½	1:2½ to 3¼	1:3 to 4
(iii)	1:1:5 to 6	1:4 to 5	1:5 to 6
(iv)	1:2:8 to 9	1:5½ to 6½	1:7 to 8
(v)	1:3:10 to 12	1:6½ to 7	1:8

← Increasing strength and improving durability

← Increasing ability to accommodate movements due to temperature and moisture changes

Increasing resistance to frost attack during construction →
Improvement in adhesion and consequent resistance to rain penetration →

*Where mortar of a given compressive strength is required by the designer, the mix proportions should be determined from tests following the recommendations of Appendix A of BS 5628: Part 1: 1978.

†The different types of mortar that comprise any one designation are approximately equivalent in compressive strength and do not generally differ greatly in their other properties. Some general differences between types of mortar are indicated by the arrows at the bottom of the table, but these differences can be reduced.

‡The range of sand contents is to allow for the effects of the differences in grading upon the properties of the mortar. In general, the lower proportion of sand applies to grade G of BS 1200, whilst the higher proportion applies to grade S of BS 1200.

§The proportions are based on dry hydrated lime. The proportion of lime by volume may be increased by up to 50% (v/v) in order to obtain workability.

‖At the discretion of the designer, air-entraining admixtures may be added to lime: sand mixes to improve their early frost resistance. (Ready-mixed lime: sand mixes may contain such admixtures.)

structural material. More detail can be found in BS 5628: Part 2: 1985.

Allowance needs to be made for the movements that occur from wetting and drying, long-term irreversible moisture expansion of clay bricks, and thermal expansion or contraction. This is done by breaking the brickwork down into defined lengths, or heights, by a complete break—a movement joint. In clay brickwork, such joints are primarily expansion joints, but in calcium silicate and concrete brickwork they are mostly for shrinkage purposes. Guidance on the spacing and detailing of movement joints is given in BS 5628: Part 3. For clay brickwork it is suggested that an allowance of 1 mm per 1 m expansion is made, so that for joints at 12 m centres 12 mm of expansion must be catered for. Since most fillers (they must be easily compressible and not, for example, cane fibre) can only accept movement of 50% of their width, this can mean rather wide joints which need to be sealed effectively. When clay brickwork is built into a framed structure (particularly, but not exclusively, reinforced concrete) joints are needed to prevent the shortening of the frame and expansion of the brickwork from causing a build-up of compressive stress in the wall.

Information on thermal resistance of brickwork may be obtained from BCR Ltd. Average sound insulation values are given in *Table 8.5*.

As for sound resistance, brick walls give good resistance to the passage of fire. The periods of resistance are given for

Table 8.5   *Average sound insulation value of brick walls*

Construction	Average sound reduction (dB)
Half-brick wall, unplastered	42
Half-brick wall, plastered both sides	45
One brick wall, plastered both sides	50
Cavity wall of two half-brick skins with butterfly wall tiles not more than one per m²; wall plastered both sides	50
One-and-a-half-brick wall	52/3

various thicknesses, plaster type and loading condition in BS 5628: Part 3: 1985.

The choice of a brick for durability in a particular building can be made from BS 5628: Part 3: 1985: Table 3 Durability of Masonry in Finished Construction; the table covers brickwork and blockwork. Frost resistance of the mortar is also necessary; if 1:1:6 or 1:$\frac{1}{4}$:3 mortars are used, this will normally be achieved. Table 3 of BS 5628: Part 3: 1985 also gives recommended mixes.

A comprehensive range of situations is covered as follows:

(1) Work below or near external ground level
(2) Damp-proof-course
(3) Unrendered external walls (other than chimneys, cappings, copings, parapets or sills)
(4) Rendered external walls
(5) Internal walls and inner leaves of cavity walls
(6) Unrendered parapets
(7) Rendered parapets
(8) Chimneys
(9) Cappings, copings and sills
(10) Freestanding boundary and screen walls
(11) Earth-retaining walls
(12) Drainage and sewerage.

Care is needed in the choice of brick and mortar for given situations, but when care is given it is repaid in durable brickwork that will last the lifetime of the building of which it forms a part.

8.2 Vitrified clay pipes for drainage and sewerage

Clays have been used as a material for making pipes for drainage since the beginning of civilization. Over 3500 years ago, clay pipes were used in the Royal Palace of Knossos in Crete—and these are still in good condition today. Vitrified clay pipes are those which have been fired to about 1100°C and which require no surface glaze.

8.2.1 Sizes

Table 8.6 indicates a range of nominal sizes from 100 to 1200 mm and corresponding minimum bores to EN 295: 1991. *Table 8.7* shows the preferred nominal lengths of pipe.

8.2.2 Standards

BS EN 295 is the European Standard for 'normal' flexibly jointed clay pipes and fittings. Part 1 of the standard specifies the product requirements, BS EN 295-1. Part 2 specifies Sampling and Quality Control, BS EN 295-2. Part 3 specifies Test Methods, BS EN 295-3. BS 65 covers unjointed normal and extra chemically resistant pipes and fittings, surface water pipes and fittings and their flexible joints, perforated pipes and fittings and ducts.

CPDA publishes a booklet 'New Standards for Vitrified Clay Pipes and Fittings' which fully describes the contents of the two standards.

Rubber seals used in jointing vitrified clay pipes are supplied to either ISO/DIS 4633 or BS 2494 (Type D) as appropriate to the standard.

Providing that pipes manufactured to BS EN 295 or BS 65 as appropriate are correctly laid and jointed, they will easily meet the test requirements of the Codes of Practice BS 8301 and BS 8005: Part 1.

Table 8.6   *EN 295-1: Minimum bore*

Nominal size (DN)	Minimum bore (mm)
100	96
150	146
200	195
225	219
250	244
300	293
350	341
400	390
450	439
500	487
600	585
700	682
800	780
1000	975
1200	1170

Table 8.7   *EN 295-1 Preferred nominal lengths*

Nominal size (DN)	Length (m)
200	1.5 2.0
225	1.5 1.75 2.0
250	1.5 2.0
300	1.5 2.0 2.5
$\geq$ 350	1.5 2.0 2.5 3.0

8.2.3 Performance of flexible joints

It is important not only that sewage or other effluents should be unable to exfiltrate pipelines, but also that ground water should be unable to enter. Flexibly jointed vitrified clay pipes to BS EN 295-1 are capable of withstanding an internal hydraulic pressure of 50 kPa (0.5 bar) for 15 minutes without leakage. Pipe joint assemblies are capable of withstanding specified angular deflection and shear resistance tests under both internal and external pressures of 5 kPa (0.05 bar) and 50 kPa (0.5 bar) for 5 minutes without visible leakage. Surface water pipes and surface water pipe joint assemblies to BS 65 are capable of withstanding an internal pressure of 30 kPa (0.3 bar) for 5 minutes.

In the straight-draw test the pipes are drawn 10 mm apart.

In the deflection test one pipe is deflected relative to the other by the value given in *Table 8.8*.

A pipe joint assembly specified to BS EN 295 is required to withstand a shear load of 25 N mm^{-1} of nominal size applied to the spigot end of one pipe while the socket of the other pipe is

Table 8.8 *Deflections of some vitrified clay pipe sizes*

Nominal size (mm)	Deflection per metre of deflected pipe length (mm)	Approximate equivalent angular deflection (°)
100–200	50	3
225–500	30	1.75
600–1000	20	1.25

held firmly in a test rig. Joints which allow more than 6 mm line displacement are supported so that this displacement is not exceeded.

The angular-deflection capabilities of the joints are important where pipes are laid in areas where movement can occur, particularly in made-up ground or an area of mining subsidence. The shear capabilities allow some differential movement between one pipe and another.

All these tests ensure that joints for vitrified clay pipes will accommodate any normal ground movement without risk of leakage.

Table 8.9 (a) *Crushing strength (FN) in kN m^{-1} DN100 and 150*

Nominal size (DN)	Crushing strength (FN)		
100	22	28	34
150	22	28	34

Table 8.9 (b) *Crushing strength (FN) in kN m^{-1} ≥DN200*

Nominal size (DN)	Class L*	Class Number			
		95	120	160	200
200			24	32	40
225			28	36	45
250			30	40	50
300			36	48	60
350			42	56	70
400		38	48	64	
450		43	54	72	
500		48	60	80	
600	48	57	72		
700	60	67	84		
800	60	76	96		
1000	60	95			
1200	60				

*Lower strength pipes

8.2.4 Crushing strength

The crushing strenghts for a range of nominal sizes of vitrified clay pipe given in BS EN 295-1 are listed in *Table 8.9* for flexibly jointed pipes.

The strengths used in design together with new bedding factors for clay pipes for classes F and B bedding recommended by the WAA are listed in *Table 8.9*; these give the limits of cover set out in *Table 8.10*, using wide trench design criteria.

8.2.5 Bending moment resistance

Minimum required bending moment resistance values to EN 295-1 and BS 65 for smaller diameter pipes are shown in *Table 8.11*.

8.2.6 Hydraulic roughness

Hydraulic roughness factors K_s as determined by HRS and WRC are shown in *Table 8.12*.

For further information the reader is referred to CPDA publications:

- Technical Notes for Vitrified Clay Pipes
- New Standards for Vitrified Clay Pipes.

8.3 Tiles and tiling

Tiles are widely used in building as a cladding, covering, or finish to roofs, walls and floors, as they have proved to provide adequately the combined functions of weather protection, resistance to wear and decoration. A number of materials may be used to form tiles, the main requirement being that the material can be formed or worked into regular thin shapes having the required properties of low permeability and attractive finished appearance. Such materials include fibre composites, thermoplastics, marble, terrazo, glass, slate, stainless steel and copper. One commonly used material is natural clay which, after processing and shaping, is baked to provide ceramic tile products.

In this section ceramic tiles for roof cladding, interior and exterior wall cladding and floor finishes are considered. Standards for the manufacture of ceramic wall and floor tiles, together with the necessary skills and details for their successful application, are fully documented in British Standards and other manuals produced by major manufacturers of tiles.

Ceramic tiles are amongst the most durable of building materials. They are resistant to attack by most acids and alkalis, and can be frost resistant. Tiling, if properly carried out, is impervious to fluids such as oils, grease and cleaning materials, and is also resistant to the weather. Clay tiles and tiling are not resistant to hydrofluoric acid or strong caustic solutions.

Wall and floor tiling acts as cladding to a construction made up from a number of layers of material of differing thickness and properties (*Figures 8.1* to *8.3*). Each layer will be subject to dimensional change depending upon its material properties and

Table 8.10 Limits of cover between which vitrified clay pipes can be laid in any width of trench

Nominal bore (mm)	Bedding construction	Crushing strength (kN m^{-1})	Main traffic roads (m)	Other roads (m)	Fields and gardens (m)
100	Class D or N	22	0.6–4.2	0.6–4.6	0.4–4.7
		28	0.4–5.7	0.5–6.0	0.4–6.0
		34	0.4–7.1	0.4–7.3	0.4–7.3
		40	0.4–8.5	0.4–8.6	0.4–8.6
	Class F	22	0.4–8.0	0.4–8.2	0.4–8.2
		28	0.4–10.0+	0.4–10.0+	0.4–10.0+
		34	0.4–10.0+	0.4–10.0+	0.4–10.0+
		40	0.4–10.0+	0.4–10.0+	0.4–10.0+
	Class B	22	0.4–10.0+	0.4–10.0+	0.4–10.0+
		28	0.4–10.0+	0.4–10.0+	0.4–10.0+
		34	0.4–10.0+	0.4–10.0+	0.4–10.0+
		40	0.4–10.0+	0.4–10.0+	0.4–10.0+
150	Class D or N	22	1.1–2.0	0.9–2.9	0.6–3.1
		28	0.7–3.4	0.7–3.9	0.6–4.0
		34	0.6–4.5	0.6–4.9	0.6–4.9
		40	0.6–5.5	0.6–5.8	0.6–5.8
	Class F	22	0.6–5.2	0.6–5.5	0.6–5.6
		28	0.6–6.9	0.6–7.1	0.6–7.1
		34	0.6–8.5	0.6–8.6	0.6–8.7
		40	0.6–10.0+	0.6–10.0+	0.6–10.0+
	Class B	22	0.6–7.2	0.6–7.3	0.6–7.4
		28	0.6–9.3	0.6–9.4	0.6–9.4
		34	0.6–10.0+	0.6–10.0+	0.6–10.0+
		40	0.6–10.0+	0.6–10.0+	0.6–10.0+
225	Class D or N	28	–	1.1–2.3	0.6–2.6
		36	0.9–2.6	0.8–3.3	0.6–3.5
		45	0.6–3.9	0.6–4.3	0.6–4.4
	Class F	28	0.6–4.3	0.6–4.7	0.6–4.8
		36	0.6–5.9	0.6–6.2	0.6–6.2
		45	0.6–7.6	0.6–7.7	0.6–7.8
	Class B	28	0.60–6.1	0.6–6.3	0.6–6.3
		36	0.6–8.0	0.6–8.2	0.6–8.2
		45	0.6–10.0+	0.6–10.0+	0.6–10.0+
300	Class D or N	36	–	1.2–2.1	0.6–2.5
		48	0.8–2.7	0.8–3.4	0.6–3.5
		60	0.6–4.0	0.6–4.4	0.6–4.5
	Class F	36	0.6–4.2	0.6–4.6	0.6–4.6
		48	0.6–6.0	0.6–6.2	0.6–6.2
		60	0.6–7.7	0.6–7.8	0.6–7.8
	Class B	36	0.6–5.9	0.6–6.1	0.6–6.2
		48	0.6–8.1	0.6–8.2	0.6–8.3
		60	0.6–10.0+	0.6–10.0+	0.6–10.0+
375	Class D or N	36	–	–	0.9–1.9
		45	–	1.1–2.4	0.6–2.7
		60	0.8–3.0	0.8–3.6	0.6–3.7
	Class F	36	0.7–3.2	0.7–3.8	0.6–3.9
		45	0.6–4.5	0.6–4.9	0.6–5.0
		60	0.6–6.4	0.6–6.6	0.6–6.7
	Class B	36	0.6–4.8	0.6–5.2	0.6–5.2
		45	0.6–6.3	0.6–6.5	0.6–6.6
		60	0.6–8.7	0.6–8.8	0.6–8.8
400	Class D or N	38	–	–	0.9–1.8
		48	–	1.2–2.3	0.6–2.6
		64	0.8–2.9	0.8–3.6	0.6–3.7

Table 8.10   *Continued*

Nominal bore (mm)	Bedding construction	Crushing strength (kN m⁻¹)	Main traffic roads (m)	Other roads (m)	Fields and gardens (m)
	Class F	38	0.8–3.0	0.7–3.7	0.6–3.8
		48	0.6–4.4	0.6–4.8	0.6–4.9
		64	0.6–6.3	0.6–6.5	0.6–6.5
	Class B	38	0.6–4.6	0.6–5.0	0.6–5.1
		48	0.6–6.2	0.6–6.4	0.6–6.5
		64	0.6–8.5	0.6–8.6	0.6–8.6
450	Class D or N	43	–	–	0.8–1.9
		54	–	1.1–2.4	0.6–2.7
		72	0.8–3.0	0.7–3.7	0.6–3.8
	Class F	43	0.7–3.2	0.7–3.8	0.6–3.9
		54	0.6–4.5	0.6–4.9	0.6–5.0
		72	0.6–6.4	0.6–6.7	0.6–6.7
	Class B	43	0.6–4.8	0.6–5.2	0.6–5.2
		54	0.6–6.3	0.6–6.6	0.6–6.6
		72	0.6–8.7	0.6–8.8	0.6–8.9
500	Class D or N	48	–	–	0.8–1.9
		60	–	1.1–2.4	0.6–2.7
		80	0.8–3.0	0.7–3.6	0.6–3.7
	Class F	48	0.7–3.2	0.7–3.8	0.6–3.9
		60	0.6–4.5	0.6–4.9	0.6–4.9
		80	0.6–6.4	0.6–6.6	0.6–6.7
	Class B	48	0.6–4.8	0.6–5.2	0.6–5.2
		60	0.6–6.3	0.6–6.5	0.6–6.6
		80	0.6–8.6	0.6–8.8	0.6–8.8
600	Class D or N	48	–	–	–
		57	–	–	0.8–1.9
		72	–	1.0–2.4	0.6–2.7
	Class F	48	1.0–2.1	0.8–3.1	0.6–3.2
		57	0.7–3.2	0.7–3.8	0.6–3.9
		72	0.6–4.6	0.6–4.9	0.6–5.0
	Class B	48	0.6–3.8	0.6–4.3	0.6–4.4
		57	0.6–4.8	0.6–5.2	0.6–5.2
		72	0.6–6.4	0.6–6.6	0.6–6.6

Table 8.11   *Bending moment resistance (BMR) in kN m⁻¹ for crushing strength values (FN) in kN m⁻¹*

Nominal size (DN)	FN	BMR	FN	BMR	FN	BMR
100	22	1,0	28	1,3	34	1,7
150	22	2,8	28	3,4	34	4,0
200	24	5,2	32	6,2	40	7,4
225	28	6,5	36	7,4	45	9,0

Table 8.12 *Design values of hydraulic roughness in used sewers for typical peak DWF velocities*

Velocity (m s⁻¹)	k_s (mm)
>1.5	0.3
>1.0	0.6
0.76–1.0	1.5

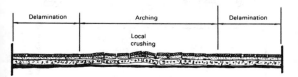

Figure 8.1 *The effect of movement on wall and floor tiling. There may be more than one zone of arching within a confined panel*

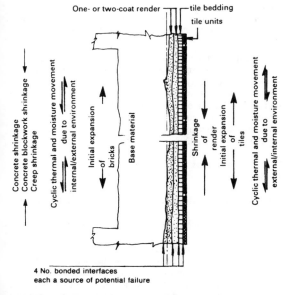

Figure 8.2 *Wall tiling (interior and external) as cladding to a construction made up of a number of layers of differing thickness and properties. Base material: in situ or precast concrete, concrete blockwork, or brickwork*

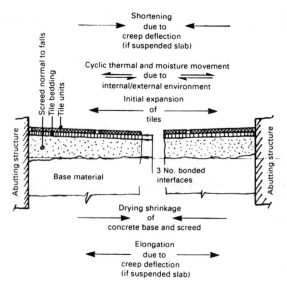

Figure 8.3 *Floor tiling ('on the ground' or suspended) as cladding to a construction made up of a number of layers of differing thicknesses and properties. Base material: usually dense concrete*

the environmental conditions imposed. In some cases adjacent layers may be straining in opposite directions leading to separation at their boundary.

In the absence of correctly designed and constructed joints to limit accumulated movement, tiling is likely to fail by delamination, arching or crushing.

Tiling is covered by the following: BSCP 5385; BSCP 202; BSCP 5534 and BS 8203.

8.3.1 Types of tiles
Roofing The two categories of tiles are:

- *interlocking tiles* which have a single lap in the longitudinal direction
- *plain tiles* which have a double lap in the lateral direction.

The origin of each category is historical, and it is easy to understand how many of the current tile profiles have been derived from the 'classical' sections illustrated in *Table 8.13*.

Plain clay tiles are also used in applications other than roofing. In certain conditions wind uplift can be produced (see *Figure 8.4*); care should be taken to adequately anchor tiles to the roof structure (see *Figure 8.5* and *Table 8.14*). For comprehensive coverage of wind pressures see BSCP 3 Chapter V: Part 2.

Table 8.13 Classical roof tiles sections

Description of tiling	Typical sections	Typical dimensions (mm)	Roof pitch (degrees to horizontal)	
			Min.	Max.
Spanish	Over tiles. Under tiles. Over tiles taper towards head, under tiles taper towards tail. 100–75. 185–145. 185–225.	Length 376	35	40–50
Italian	Over tiles. Under tiles. 75. 150. 235.	Length 375	35	40–50
Single Roman	Over and under tiles taper towards head and tail esp. 75. 250.	250 × 340	35	40–50

Table 8.13 Continued

Description of tiling	Typical sections	Typical dimensions (mm)	Roof pitch (degrees to horizontal)	
			Min.	Max.
Double Roman	75 · 345	340 × 420	35	40–50
Pantiling	60 · 250	250 × 335	30	47.5
Plain	Nibs · Long-section · Nibs · Cross-section	165 × 265	40	Vertical

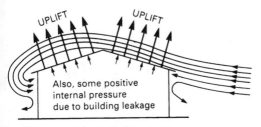

Figure 8.4   *Generation of uplift on shallow pitch roofs*

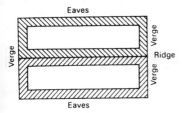

Figure 8.5   *Plan of typical roof showing ridge, verge and eaves bands for minimum fixings. Width of bands is approximately three tiles*

Wall tiles   Wall tiles are typically referred to as ceramic tiles, unlike roof tiles which retain their clay designation. Wall tiles for internal and external use may be glazed or unglazed, plain or decorated. The surface of a tile may be smooth, textured or profiled, according to its function or intended finished appearance. Tiles for external use must be frost resistant, and this property should be verified with the manufacturer.

Although other shapes are available, tiles manufactured for internal use are normally either square or rectangular, with face dimensions ranging from 300 mm to 100 mm, with thickness from 8.5 mm to 5.5 mm. In addition, for external use, larger sizes are available, up to $600 \times 300 \times 30$ mm.

8.3.2 Properties
Typical mechanical and physical properties are shown in *Table 8.15*. Permissible deviations are shown in *Table 8.16*.

8.3.3 Mosaics
Mosaics consist of coloured marble or ceramic prisms, from 6 mm^2 to 25 mm^2 square or hexagonal shapes on plan and 6 mm to 25 mm thick. The tesserae are glued to a piece of paper on which a pattern or picture is often drawn, and then reversed for fixing. Sometimes nylon net or strips are glued to the back face with the intention of enhancing adhesion with the

Table 8.14 *Example of how the fixing pattern for duo-pitched roofs varies with roof pitch and wind speed: two-storey house, height to ridge of structure H = 6.6 to 9.2 m single lap interlocking tiles only*

	Basic wind speed V = 38 m s⁻¹				Basic wind speed V = 50 m s⁻¹			
	Ground roughness category				Ground roughness category			
Pitch (°)	1	2	3	4	1	2	3	4
17.5–22								
22.5–29.5								
30–44.5								
≥ 45								

	Key	Wind uplift (kN m^{-2})	Suggested mechanical fixings
(i)		0–0.07	No fixing required for uplift, but advisable to use BS 5534: Part 1: 1978, CL 34.4 minimum (i.e. every perimeter tile should be mechanically fixed).
(ii)		0–0.7	No fixing required for uplift, but in every tile one or two nails at head (to reduce chatter and shear on pitches of 45, 55, over 55 use (xi))
(iii)		>1.0	Use verge clips at verge perimeter, or preferably dry or cloaked verges.
(iv)		0.7–1.2	Every tile one nail at head (most convenient hole).
(v)		1.2–1.5	Every tile two nails at head (if risk of cracking tile use next fixing key).
(vi)		0.7–1.2	Alternative tiles in each row (chequer pattern) clipped at toe every perimeter tile clipped. Note that nail holes are not practical, because of risk of water penetration, at pitches of less than 30.
(vii)		1.2–1.5	
(viii)		1.5–2.0	Alternative tiles in each row (chequer pattern) clipped at toe plus every perimeter tile clipped.
(ix)		1.5–2.0	Combination of (viii) and (iv) (to reduce chatter and shear on pitches over 45, 55, over 55 use (xi)).
(x)		2.0–2.5	Every tile one clip at toe.
(xi)		2.5–3.1	Every tile one clip at toe and one nail at head.
(xii)		2.5–3.1	Every tile one clip at toe. Note nail holes are not practical, because of water penetration, at pitches of less than 30. This is a reduction in fixing and slightly increases the risk of uplift.

NB. Fixing each tile in alternate rows is *not* recommended, being much less effective than alternate tiles in each row (chequer pattern).

Table 8.15 *Typical values of the principal mechanical and physical properties of tiles*

Property	Wall tiles, dust pressed	Floor tiles, extruded		Roof tiles, plain clay
		Split	Quarries	
Modulus of rupture ($N mm^{-2}$)	27	20	20	15
Coefficient of thermal expansion ($\times 10^{-6} K^{-1}$)	9	4–8	5–13	8
Moisture movement		$0.6 mm m^{-1}$ for tiles where $6\% < E \leq 10\%$		

bedding mortar or adhesive. Mosaics are supplied in square or rectangular sheets suited to handling, which restricts the maximum dimension to about 600 mm.

Mosaic facades or floor coverings will not be successful unless used in conjunction with the very highest quality of workmanship. Unless this can be provided the use of mosaics is not recommended.

Bibliography

DORAN, D.K. (ed.), *Construction Materials Reference Book*, Butterworth-Heinemann, Oxford (1992):

Chapter 11: Haseltine, B.A.
Chapter 12: Bland, C.E.G.
Chapter 13: Baker, R.C.

(1) Contamination by evaporite minerals, most notably chlorides and sulphates
(2) Reactive minerals within available materials, with the potential for alkali-silica, alkali-silicate and alkali-carbonate reactivity
(3) Materials which are unsound as opposed to reactive or whose physical breakdown may encourage reactivity.

Although each has been identified in the Middle East, rapid chloride-promoted rusting of reinforcement in concrete has tended to mask these longer-term processes of deterioration.

Active wadis within the mountains and a limited portion of the alluvial fans formed where they emerge onto the desert plain are sources of relatively clean, well-graded sands and gravels. Their torrent-bedded, poorly sorted structure makes them potentially good sources of 'all-in' material. The active periodic flooding ensures that groundwater movement is potentially downwards and that sulphates and chlorides do not accumulate to significant levels.

Upward movement of groundwater is, however, the dominant process and gives rise to a hard, well cemented surface layer called a 'duricrust', composed of precipitates from pore fluids. This layer most commonly comprises calcium and calcium-magnesium carbonates (calcrete and dolocrete), calcium sulphate (gypcrete if strong, gypcrust if weak), and sodium chloride (salcrust). When the soluble materials are inert, the improvement in material properties may enable them to be used as aggregate sources. This is most commonly the case where weak limestones develop a calcrete capping. The accumulation of sulphates and chlorides more commonly render this an adverse process, although calcrete has no definite base and passes gradually into unhardened limestone below. The depth of working must be carefully controlled in order to maintain quality. However, calcretes generally cannot be considered as particularly good materials on physical and mechanical grounds, apart from possible chemical problems.

Silica sands are rare in the Persian Gulf region where carbonate sands predominate. Also, the heavy dependence on beach and coastal sands as sources of fine aggregate in the Middle East reflects the poor gradings of inland deposits.

Contamination of stockpiles of aggregate in the Middle East must be prevented by placement on a layer of dense concrete or asphalt laid to a fall and provided with effective drainage. Windbreaks may also be required to prevent wind-blown dust or salts contaminating stockpiles.

9.3.3.3 Types of aggregates for concrete The classification of natural aggregates is as lightweight or dense according to their bulk density or particle density. In the BSI system, a fine aggregate is one in which particles mainly pass a 5.0 mm BS 410: 1983 test sieve and does not contain more coarse material than permitted for the grading specified. A coarse aggregate is one in which most particles are retained on a 5.0 mm BS 410: 1983 test sieve with no finer material than permitted for the various sizes specified.

In the ASTM system, 4.75 mm (No. 4 sieve) is the nominal dividing size of coarse and fine aggregate. The system under the DIN scheme relates mainly to all-in aggregate, the grading characteristics being specified according to the maximum size of aggregate particles.

Coarse aggregate can be:

(1) Gravel: (i) uncrushed gravel which results from the natural disintegration of rock; (ii) crushed gravel; or (iii) partially crushed gravel
(2) Crushed rock
(3) Blended coarse aggregate produced by the controlled blending of gravel and crushed rock.

Fine aggregate can be:
(1) Sand: (i) uncrushed sand; or (ii) partially crushed sand
(2) Crushed gravel fines
(3) Crushed rock fines
(4) Blended fine aggregate.

In addition, 'all-in aggregate' may be produced which is a mixture of both coarse and fine fractions.

Of particular interest when selecting aggregates is the extent to which they will react chemically with other mix constituents. It is now widely accepted that all aggregates react with cement. With some the reaction is negligible whilst others react strongly producing the more damaging effects of ASR. *Table 9.11* lists those aggregates whose reaction is considered sufficiently low to be classified as non-reactive for all practical purposes. For those aggregate or rock types not included in *Table 9.10* guidance is given in *Table 9.12*.

Table 9.11 *Components of aggregate considered non-alkali silica reactive*

Air-cooled	Expanded	Microgranite
blast-furnace	clay shale slate	Quartz*†
slag	Feldspar*	Schist
Andesite	Gabbro	Sintered pulverized-fuel ash
Basalt	Gneiss	Slate
Diorite	Granite	Syenite
Dolerite	Limestone	Trachyte
Dolomite	Marble	Tuff

*Feldspar and quartz are not rock types but are discrete mineral grains principally in fine aggregate
†No highly strained quartz and not quartzite.

Table 9.12 *Some guidance for the assessment of rock types not included in Table 9.10*

Rock type	Definition	Potentially alkali-reactive components that may sometimes be present
Arkose	Detrital sedimentary rock containing more than 25% feldspar	See sandstone
Breccia	Coarse detrital rock containing angular fragments	See sandstone
Chert	Micro- or crypto-crystalline silica	See flint
Conglomerate	Coarse detrital rock containing rounded fragments	See sandstone
Flint	Strictly, chert occurring in cretaceous chalk	Chalcedonic silica and micro- or crypto-crystalline quartz. Some varieties may contain opaline silica
Granulite	Metamorphic rock	Highly-strained quartz
Greywacke	Detrital sedimentary rock containing poorly sorted rock fragments and mineral grains	May be alkali-silicate reactive.* See sandstone
Gritstone	Sandstone with coarse, angular grains	See sandstone
Hornfels	Fine-grained, thermally metamorphosed rock	Glass‡ or devitrified glass. Highly-strained‡ and/or microcrystalline quartz. Phyllosilicates*
Quartz	Discrete mineral grains very common in fine aggregates	Highly-strained‡ quartz
Quartzite	(i) Sedimentary or *ortho*-quartzite (ii) Metamorphic or *meta*-quartzite	See sandstone. Highly-strained‡ quartz and/or high-energy quartzite grain boundaries
Rhyolite	Fine-grained to glassy acid volcanic rock	Glass‡ or devitrified glass. Tridymite. Cristobalite. Opaline or chalcedonic veination or vugh-fillings
Sandstone	Detrital sedimentary rock. The grains are most commonly quartz, but fragments or grains of almost any type of rock or mineral are possible	Highly-strained quartz. Some types of rock cement, notably opaline silica, chalcedonic silica, and micro-crystalline or crypto-crystalline quartz. Phyllosilicates

*Phyllosilicates are sheet silicate minerals, including the chlorite, vermiculite, mica and clay mineral groups. Within the UK a few cases of possible alkali silicate reaction have been reported in coarse aggregates containing greywacke and related rocks. The matrix in such rocks is very finely divided and consists of phyllosilicates, quartz and other minerals.
†Rocks containing more than 5% (by volume) glass, partially devitrified glass or devitrified glass should be classified as potentially alkali reactive.
‡If the average undulatory extinction angle obtained from at least 20 separate quartz grains (measured in thin section under a petrological microscope) is more than 25° the quartz should be classified as 'highly strained'. Rocks containing more than 30% highly-strained quartz should be classified as potentially alkali reactive.

9.3.3.4 Impurities In order to produce durable concrete it is essential that the level of impurities in aggregates is kept at safe limits. Such impurities include:

- chalk — often frost susceptible and may cause 'pop-outs'
- coal — may be frost susceptible, cause 'pop-outs' and surface staining
- sulphides — pyrites and marcasite may cause surface staining
- micas — weak flaky materials which may require higher water : cement ratios to maintain desirable workability
- clay, silt, dust — these materials may coat the aggregates thus reducing bond with cement paste
- shell — may not be harmful in small quantities but good knowledge of grading and chemical content (particularly sulphates and chlorides) is essential before permitting use
- chlorides and sulphates — both potentially damaging ingredients. Chlorides cause corrosion of steel reinforcement. Guidance on safe limits given in *Tables 9.13* and *9.14*.

9.3.3.5 Physical properties of aggregates The correct selection and subsequent control of aggregates may require a programme of testing. Many such tests are available and guidance to some is given in *Tables 9.15* and *9.16*. Arguably the key

Table 9.13 *Chloride content of concretes from marine aggregates*

Concrete grade	C30	C35	C40	C45	C50
Minimum cement content ($kg\,m^{-3}$) BS 8110: 1985 (*Table 9.4*)	275	300	325	350	400
Assumed maximum aggregate : cement ratio	6.9	6.3	5.7	5.3	4.5
Chloride ion in combined aggregate (wt%)					
Maximum chloride ion in concrete (wt% cement)*	0.06	0.46	0.43	0.39	0.37	0.32
	0.05	0.40	0.37	0.34	0.32	0.28
	0.04	0.33	0.30	0.28	0.26	0.23
	0.03	0.26	0.24	0.22	0.21	0.19
	0.02	0.19	0.18	0.17	0.16	0.14
	0.01		0.11	0.11	0.10	0.10

*Derived from aggregate + cement. (The chloride content of cement is assumed to be at the maximum of the typical range of 0.01–0.05% for ordinary Portland cement.)

Table 9.14 *Suggested working limits for chloride and sulphate contamination in concrete aggregates in the Middle East*

Total chlorides (as Cl)	In fine aggregate not more than 0.06% Cl by weight of fine aggregate
	In coarse aggregate not more than 0.03% Cl by weight of coarse aggregate
	Overriding requirement: in concrete not more than 0.3% Cl by weight of cement
Total sulphates (as SO_3)	In fine aggregate not more than 0.4% of fine aggregate
	In coarse aggregate not more than 0.4% of coarse aggregate
	Overriding requirement: in concrete not more than 4.0% by weight of cement

characteristic of an aggregate to produce optimum levels of workability and strength is grading. The most desirable features of grading are:

(1) the lowest possible surface area per unit volume; and
(2) the lowest void content per unit volume; resulting in
(3) the lowest possible water content; and
(4) the lowest cement content.

It should be noted that special precautions are necessary in regions such as the Middle East where the risk of contamination is high and the environment potentially damaging.

9.3.3.6 Lightweight aggregates Lightweight aggregates are specialized materials having an apparent specific gravity significantly less than that of normal fine and coarse mineral aggregates. They range from extremely lightweight types used in insulating and non-structural concrete, to expanded shales, clays and slates used in structural concrete. The apparent specific gravity depends on the quantity of air contained, the highest air contents enhancing insulation but producing lower strengths. British Standards define lightweight aggregate as having a density of less than $960 \, \text{kg m}^{-3}$ for coarse aggregate or $1200 \, \text{kg m}^{-3}$ for fine aggregate: alternatively, having a particle density of not more than $2000 \, \text{kg m}^{-3}$. The ASTM definition is of 'aggregate of low density used to produce lightweight concrete, including: pumice, scoria, volcanic cinders, tuff, and diatomite; expanded or sintered clay, slate, diatomaceous shale, perlite, vermiculite, or slag; and end products of coal or coke combustion'.

The main applications of lightweight concrete in the UK have been in the form of block masonry. However, some structural applications, notably on prestressed lightweight concrete in grandstands, have been employed. Lightweight aggregate concrete is dealt with for structural uses in both BS 8110: 1985 and the Code of Practice for Design of Concrete Bridges. However,

Table 9.15 Aggregate tests by standard and specification

Test	Standard			
	BS	ASTM	DIN	Other (ref. no.)
Definitions and descriptions	BS 6100: 1984, Section 5.2	C294-86, C638-84	4226: Part 1	
Grading, dry method	BS 812: 1985, Part 103	C136-84a	4226: Part 3, 3.1	
Grading, wet method	BS 812: 1985, Part 103	C117-87	4226: Part 3, 3.1	
Clay, silt, dust content	BS 812: 1975, Part 1	C117-87, C142-78	4226: Part 3, 3.6.1	Afnor, 1980
Relative density and water absorption	BS 812: 1975, Part 2, Clause 5	C127-84	4226: Part 3, 3.4	
Moisture content		C128-84		
Surface moisture		C556-84		
Particle shape	BS 812: 1985, Section 105.1 BS 812: 1975, Part 1, Clause 7.4	D3398-81	4226: Part 3, 3.2	
Bulk density of aggregates including voids and bulking	BS 812: 1975, Part 2, Clause 6.0	C29-87	4226: Part 3, 3.3	
Lightweight pieces		C123-83		
10% Fines value	BS 812: 1975, Part 3, Clause 8			
Aggregate impact value	BS 812: 1975, Part 3, Clause 6			
Aggregate crushing value	BS 812: 1975, Part 3, Clause 7			
Aggregate abrasion value	BS 812: 1975, Part 3, Clause 9			
Polished stone value	BS 812: 1975, Part 3, Clause 10			
Los Angeles abrasion value		C131-81, C535-81	52, 100	
Petrography	BS 812: Part 104 (draft)	C295-85		

	ASTM		BS 812	BS 4226	References
Soundness/frost resistance	C88-83, C33-81,			4226: Part 3, 3.5	
Potential alkali-aggregate reactivity	C682-87			4226: Part 3, 4.2	
Mortar bar	C227-87				
Chemical methods	C289-87	38			
In presence of admixtures	C441-81				
Rock cylinder method	C586-81				
Osmotic cell test					Stark, 1983
Mortar prism test					Chatterji, 1978; Oberholster and Davies, 1986; Ming-Shu, 1983; Tamura, 1984
					CSA, 1977; BSI, 1988; Hobbs, 1985
Concrete prism test					Jones and Tarleton, 1958
					BRE Digest 35 (1968)
Gel pat test					
Drying shrinkage					
Organic content	C40-84			4226: Part 3, 3.6.2	
Sulphate content			BS 812: 1988, Part 118	4226: Part 3, 3.6.4	
Chloride content			BS 812: 1988, Part 117	4226: Part 3, 3.6.5	
Impurities affecting hardening				4226: Part 3, 3.6.3	

Table 9.16 *Suggested schedule of tests on Middle East aggregates for use in concrete under exposed conditions*

Scope	Test	Authority	Suggested limits	Remarks
Physical properties and classification	Grading	BS 812: Part 103	BS 882	
	Elongation index	BS 812: Part 103	Not exceeding 25%	
	Flakiness index	BS 812: Parts 1 & 105.1	Not exceeding 25%	
	Specific gravity	BS 812: Part 2	–	Limits dependent on rock type.
	Water absorption	BS 812: Part 2	Not exceeding 20%	
	Soundness	ASTM C-88	Loss not exceeding 16%	Five cycles using magnesium sulphate solution
	Aggregate shrinkage	BRE Digest No. 35	Not exceeding 0.05%	
	Petrography	ASTM C-295		
Contamination and reactivity	Silt, clay and dust	BS 812: Part 1	BS 882	
	Clay lumps	ASTM C-142	Not exceeding 2.0%	
	Organic impurities	ASTM C-40	Advice given in ASTM C-40	
	Sulphate content	BS 1377: Test 9	Not exceeding 0.4% (w/w)	Subject to overall limits on total mix
	Chloride content	BS 812: Part 117	Not exceeding 0.06% (w/w)	
	Potential alkali reactivity	ASTM C-227 or C-289	Advice given in ASTM C-227 and ASTM C-289	
Mechanical properties	10% fines value	BS 812: Part 3	Not less than 8 ton	
	Aggregate crushing value	BS 812: Part 3	Under 25%	Officially superseded by 10% fines test but still widely used
	Aggregate impact value	BS 812: Part 3	Not exceeding 22%	
	Los Angeles abrasion	ASTM C-131 or C-535	Not exceeding 40%	Relevant to wearing surfaces only

its use may be restricted by imposed reductions in the design stresses in shear and torsion, the permissible span: depth ratios, the modulus of elasticity, and also by increases in anchorage lengths. The lower stresses from self-weight do, however, partially offset these effects.

A recent study made by the Concrete Society on the economics of the use of lightweight aggregate concrete for bridges considered structures with a span of about 25 m where lightweight aggregate concrete had been used for precast beams and the *in situ* deck, but not for columns or abutments. Typically, five beams rather than six could be used because of the weight saving, although the decks were thicker because of higher transverse bending movements resulting from wider beam spacing.

Small cost savings accrued overall from using lightweight concrete in such circumstances, with further savings arising from reduced foundation sizes, reduced column sizes and formwork requirements, generally lighter formwork and falsework and lower erection loads. Savings were quantified as 4% for 60 m spans and 8% for 200 m spans.

Lightweight aggregates can, in the most general sense, be classified as either suitable or unsuitable for use in structural concretes. However, an alternative view is that all concretes of a particular strength should be considered in the same way, irrespective of the aggregate used. Recent research on shear strength of lightweight aggregate tends to support the latter view.

In addition to natural sources of lightweight aggregate the following represent the more commonly available manufactured types.

Expanded clays, shales and slates   Lightweight aggregate from shale and slate involves their controlled heating by the rotary kiln of the sintering method. Siliceous clays, shales and slates will bloat on heating as a result of the liberation of gases at the temperature of incipient fusion. The resulting vitrified material is crushed and graded after cooling. Typical proprietary products are Aglite, Leca and Sintag, each manufactured from clay and shale. For environmental reasons slate is no longer used in the UK.

Expanded slag   The raw material for this is molten slag from pig iron furnaces. When controlled amounts of water or steam with compressed air is applied to the molten slag a porous material known as 'foamed slag' is produced. The additional process of bloating with water in a rotating drum produces rounded 'pelletized expanded slag'. Proprietary versions of this material are sold as Pellite or Lycrete. This material is covered by BS 877: 1977.

Furnace clinker, ash and slag   Due to a change to oil, gas and pulverized fuels the availability of furnace clinker has declined. The material is dealt with in BS 1165: 1977.

Furnace slag may be used although its sensitivity to volume change in the presence of moisture is hazardous when combined with OPC. This hazard is reduced when used in conjunction with pulverized-fuel ash.

Sintered pulverized-fuel ash   The fuel gases of modern thermal power stations produce pulverized-fuel ash. Such material when wetted, mixed with coal slurry, pelletized and sintered produces a lightweight aggregate sold commercially as Lytag.

Exfoliated vermiculite   Vermiculites (see Chapter 20) expand considerably when heated. This forms a material which is suitable for a limited range of concretes with good insulation properties but low strength.

Expanded perlite   This is a glassy volcanic rock which when heated to high temperatures will produce a cellular material suitable for use as a lightweight aggregate.

Plastic   For very low density concrete it is possible to use resin-cement coated plastic beads (e.g. expanded polystyrene) as an aggregate.

Properties   Air-dry densities, lightweight aggregate tests and typical gradings are shown in *Tables 9.17, 9.18* and *9.19* respectively.

9.3.3.7 Admixtures   Admixtures and polymers have been incorporated into concretes and mortars, knowingly or unknowingly, since ancient times. The most widely documented evidence of this is possibly the practice of Roman engineers of adding ox blood to building mortars in order to improve the physical properties of such mortars.

Since those times, the use of admixtures has become universal. In the USA, Japan, Germany, Australia and the CIS, the acceptance is so wide that virtually all concretes contain an admixture.

Table 9.17   *Air-dry densities (loose) of lightweight aggregates*

Type	Density ($kg\,m^{-3}$)
Sintered pulverized-fuel ash	770–1040
Expanded slag	700–970
Expanded clays and shales	320–960
Pumice	480–880
Diatomite	450–800
Expanded perlite	50–240
Exfoliated vermiculite	60–160
Plastic	10–20

Table 9.18 *Lightweight aggregate tests (in addition to Table 9.15) by standard*

Material	Test	Standard		
		BS	ASTM	DIN
Clinker and furnace bottom ash	All physical and chemical	BS 1165: 1985 BS 3681: Part 2: 1983		
Lightweight aggregates	All physical and chemical	BS 3681: 1983, Part 2	C330-87, C331-87	4226, Part 2 4226, Parts 3, 6 and 7
	Organic impurities		C332-87	
	Staining		C40-84	
	Loss on ignition		C641-82 C114-85	
Blast-furnace slag	Description			4226, Part 3, 5.1
	Lime unsoundness	BS 1047: 1983		4226, Part 3, 5.2.1
	Iron unsoundness	BS 1047: 1983		4226, Part 3, 5.2.2
	Dicalcium silicate unsoundness (falling)			

Table 9.19 *Typical gradings for use in lightweight concrete*

Type	Typically available gradings (mm)
Sintered pulverized fuel-ash	12-8, 8-5, 5 down (crushed)
Expanded slag	
Foamed	14-3, 3 down
Pelletized	12-3, 3 down
Expanded clays and shales	
Aglite	15-10, 10-5, 5 down
Leca	20-10, 10-3, 3 down
Sintag	14-5, 5 down
Expanded perlite	6 mm max. (Zone L1 of BS 3797)
Exfoliated vermiculite	7-6, 6-5
Plastic	4 (max.)

The United Kingdom is not typical of modern western
Europe, and admixture usage has been slow to gain the degree
of acceptance that would have been expected. There are a
number of reasons for this, mainly historical in origin. The
early admixture industry built up a poor image, based upon
products of variable quality, for which exaggerated technical
claims were made. The problems of corrosion associated with
the uncontrolled use of calcium chloride based admixtures
further damaged the reputation of the admixture industry.
Finally, the failure of some structures utilizing high alumina
cement based concretes caused many engineers to be very
wary of any untried materials and the construction industry
in the UK became deeply conservative.

It was the emergence of larger and more professional
admixture companies, who were able to offer higher quality
products, and a higher level of service that enabled the admixture industry to overcome its poor, past performance. Largely
through the initiative of such a new and professional industry,
national standards were introduced for concrete admixtures
during the 1970s which did much to redress the past, and
overcome the conservatism which had arisen out of those
technical shortcomings.

The formation of the Cement Admixtures Association (CAA)
has provided further impetus to creating an improved image of
the admixture industry. To this end, the introduction of the
CAA Quality Scheme has enabled the UK industry to achieve
a deserved credibility.

The growth in the use of admixtures has not resulted solely
from the effects of the admixture industry itself. An increased
awareness of the technical properties of concrete, combined
with an appreciation of the commercial opportunities and the
practical limitations of the material, have led to a greater understanding and confidence within the construction industry.
Cement replacement materials such as ground granulated
blast-furnace slag, pulverized fuel ash and silica fume have all
gained acceptance, and the use of admixtures in conjunction
with such materials is now widely accepted.

This section summarizes the main groups of admixture. The main types of admixture covered are air-entraining agents and mortar plasticizers, water-reducing agents and concrete plasticizers, superplasticizers, accelerators, corrosion inhibitors, retarders and permeability reducers.

Air entraining agents Air entrainment is the deliberate inclusion in concrete or mortar of a small quantity of air in uniformly distributed, very small individual bubbles. This technique improves the durability of concrete and in particular gives increased resistance to freeze–thaw damage. The use of air entrainment was first noted in the USA in the 1930s in the production of concrete used for roads. *Figure 9.6* indicates some of the benefit of using air entrainment.

There are several types of air entrainment agents, the mechanism of air entrainment varying with type of agent. The most widely used agent is derived from the alkali salt of wood resin (sodium abietate) usually extracted from pinewood. One such agent is available under the proprietary name of Vinsol. Sodium abietate is one type of anionic surfactant—others include sodium dodecyl sulphonate and sodium deconoate.

To achieve optimum benefit for durability it is essential to entrain the correct level of air into the cement paste. As the maximum aggregate size increases so the paste volume

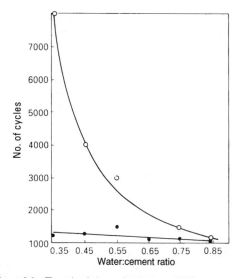

Figure 9.6 *The role of air entrainment in reducing frost damage in concrete. Water:cement ratio of concrete moist cured for 14 days versus number of cycles needed to cause 25% loss in weight, ○, Air entrained; ●, control, no air*

Table 9.20   *USA and UK recommended levels of air entrainment*

Nominal maximum aggregate size (mm)	Average air content (%)	
	UK	USA
40	4.0	5.5
20	5.0	6.0
10	7.0	7.5

decreases, thus establishing a relationship between required entrainment and aggregate size. UK and USA recommended levels of entrainment are shown in *Table 9.20*. Additional factors influencing entrainment are highlighted in *Table 9.21*.

Air entrainment and cement substitutes   The production and testing of trial mixes is recommended.

Ground granulated blast furnace slag of normal degrees of fineness has minimal effect on the required level of air entrainment.

The use of air entrainment agents with UK pulverized-fuel ash invariably requires higher than normal dosages.

Because of its very fine nature silica fume usually requires higher than normal dosages of air entraining agents.

Effect on strength   For low cement content concretes where harsh or poorly graded aggregates are used strengths may actually increase. For concretes with high cement content strengths will usually decrease unless attempts are made to reduce the quantity of fines. As with the use of cement substitutes a thorough programme of trial mixes with appropriate testing is recommended before finalizing a choice. An indication of the variation of strength in relation to cement content is shown in *Table 9.22*.

Standard specifications   The general standard for air entraining agents is BS 5075: Part 2: 1982. Methods for measuring the air content (pressure method, volumetric method and gravimetric method) are described in BS 1881: Part 106: 1983 and ASTM C231-86, ASTM C173-78 and ASTM C138-81 respectively. Additional information can be obtained from other ASTM publications. The response of concretes to BS 5075 freeze-thaw cycling is shown in *Figure 9.7*.

Mortar plasticizers (mortar air entrainers)   Mortar plasticizers, unlike concrete plasticizers, are air entraining agents, and as such they are generically similar to those used in concrete. Their main function is to improve the plasticity and rheology of building mortars. Traditional plasticizers are based on dilute solutions of neutralized wood resin but specialist products

Table 9.21 Factors that influence the air content of concrete[a]

Increasing air content	Decreasing air content	Example change	Estimated effect (target 5% air content)
Lower temperature	Higher temperature	10–20°C	Reduction 1–1.25%
Higher slump	Lower slump	50–100 mm	Increase of 1%
Sand grading coarser	Sand grading finer	Fine to medium, or medium to coarse	Increase of <0.5%
Sand content increased	Sand content decreased	35–45%	Increase 1–1.5%
Decrease in sand fraction passing 150 μm	Increase in sand fraction passing 150 μm	+50 kg m^{-3}	Reduction of 0.5%
Decrease in cement content inclusive of sand-content adjustment	Increase in cement content inclusive of sand-content adjustment	+50 kg m^{-3}	Reduction of 0.5%
–	Inclusion of organic impurities	Inclusion	Positive and negative effects reported
–	Inclusion of PFA	Inclusion	Significant reduction linked to carbon in ash
–	Increase in hardness of water	Increased hardness	Reduction
Increase in mixing efficiency	Decrease in mixing efficiency	Better mixing efficiency	Increase linked to dispersion of admixture
Positive dispensing tolerance	Negative dispensing tolerance	±5%	±0.25%
–	Prolonged agitation	1 h	Reduction of 0–0.25%
		2 h	Reduction of 1%

[a]The values given are indicative only. The effects should not be treated cumulatively since they are not necessarily independent.

Table 9.22 The effect of air entrainment on concrete strengths

Cement content ($kg\,m^{-3}$)	Change in flexural strength (%)	Change in compressive strength (%)
225	+2	+5
310	−13	−19
390	−11	−21

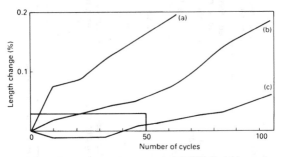

Figure 9.7 The response of concretes to BS 5075: Part 2 freeze-thaw cycling (24 h cycle): (a) no air entrainment; (b) poor air entrainment; (c) good air entrainment

relating to retarded mortars have now emerged. Products in this area are embraced by BS 4887: Parts 1 and 2: 1986.

Water-reducing agents and plasticizers for concrete Concrete plasticizers are a group of materials which, when added to concrete in small dosages, are able to impart significant increases in workability. Alternatively, the same compounds can be added to concrete, and, in order to maintain a constant workability, a water reduction can be made. Concrete plasticizers should not be confused with mortar plasticizers which are chemically dissimilar.

In the 1930s organic compounds were used in the USA to improve the workability of concrete. In the 1950s the use of hydroxy materials increased. Today many chemicals are used to give specific properties to concrete. These include lignosulphonates, hydroxylated polymers, hydroxy carboxylic acids and acrylic acid/acrylic ester copolymers. The addition of a plasticizer to a mix will result in an increase in workability; it will also increase the time over which the mix remains in a workable state. This feature is illustrated in *Figure 9.8*. Where a plasticizer/water reducing agent is used and the water content is reduced to retain the original workability, then the concrete attains a lower water : cement ratio whilst developing a higher strength (see *Figure 9.9*). Similarly plasticizers can be used to achieve a given grade of concrete at lower cement levels (see *Figure 9.10*).

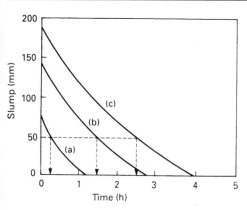

Figure 9.8 *The principle of workability extension using a concrete plasticizer. (a) Control concrete, no admixture; (b) single dose of lignosulphonate plasticizer; (c) double dose of lignosulphonate plasticizer*

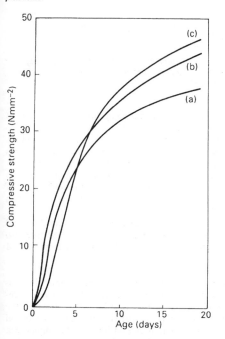

Figure 9.9 *Compressive strength gain (at constant workability) using water-reducing agents. (a) Control concrete, water : cement = 0.55; (b) plasticized/water-reduced concrete, water : cement = 0.50; (c) water-reduced/retarded concrete, water : cement = 0.50*

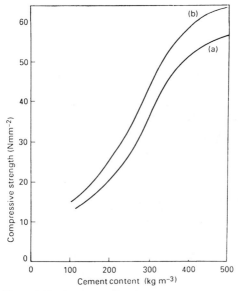

Figure 9.10 *Cement content related to the compressive strength for plain and plasticized mixes. (a) Control concrete, no admixture; (b) water-reduced concrete*

The performance of water-reducing agents varies with cement type. It is therefore advisable to seek specialist advice or to carry out trial mixes before use of any particular product. Similar caution should be applied in cases where cement has been partially replaced by ggbfs, pfa, silica fume or other products. Some guidance is given in relation to strength gain in *Figures 9.11* and *9.12*. Because it is extremely fine and may itself contain an admixture particular care should be exercised when using silica fume.

The British Standard covering accelerating, retarding and water-reducing admixtures is BS 5075: Part 1: 1982 (see also *Table 9.23*). Further guidance may be obtained from ASTM C494.

Superplasticizers These are similar materials to normal plasticizers but have the advantage of causing few adverse side-effects. This is illustrated in *Figures 9.13* to *9.15* in which the use of a superplasticizer is compared with that of a normal lignosulphonate plasticizer. It should however be emphasized that relatively high dosages of superplasticizers are required to produce the desired effect.

Superplasticizers may be used to considerably increase the workability of the concrete, even to the extent of producing

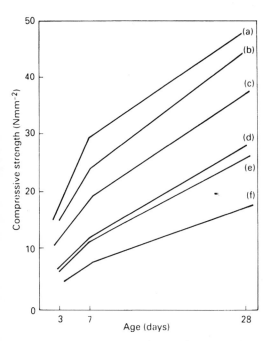

Figure 9.11 The strength gain of mixes containing 40% ground-granulated blast-furnace slag, 60% ordinary Portland cement, and plasticizing admixtures: (a) 300 grade, hydroxy carboxylic acid; (b) 300 grade, lignosulphonate; (c) 300 grade, control; (d) 200 grade, hydroxy carboxylic acid; (e) 200 grade, lignosulphonate; (f) 200 grade, control

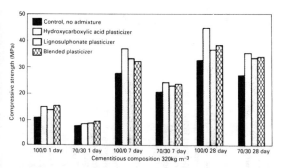

Figure 9.12 The strength gain of mixes containing pulverized fuel ash with plasticizing admixtures: all concretes to 50 mm slump

Table 9.23 The performance requirements and tests stated in BS 5075: Part 1: 1982

Property	Type of admixture				
	Accelerating	Retarding	Normal water-reducing	Accelerating water-reducing	Retarding water-reducing
Test-mix concrete A					
Compacting factor relative to control-mix concrete	Not more than 0.02 below	Not more than 0.02 below	At least 0.03 above	At least 0.03 above	At least 0.03 above
Stiffening times for 0.5 N mm^{-2}	More than 1 h	At least 1 h longer than control mix	–	–	–
for 3.5 N mm^{-2}	At least 1 h less than control mix	–	–	–	–
Minimum compressive strength as % of control-mix concrete					
at 24 h	125	–	–	125	–
at 7 days	–	90	90	–	90
at 28 days	95	95	90	90	90
Test-mix concrete B					
Compacting factor relative to control-mix concrete	–	–	Not more than 0.02 below	Not more than 0.02 below	Not more than 0.02 below

Stiffening time				
for 0.5 N mm^{-2}	–	Within 1 h of control mix	More than 1 h	At least 1 h longer than control mix
for 3.5 N mm^{-2}	–	Within 1 h of control mix	At least 1 h less than control mix	–
Minimum compressive strength as % of control-mix concrete				
at 24 h	–	–	125	–
at 7 days	–	110	–	110
at 28 days	–	110	110	110

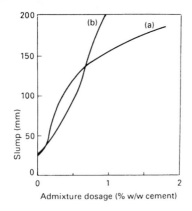

Figure 9.13 *A comparison of the effects on workability of the addition of: (a) a normal lignosulphonate plasticizer, and (b) a superplasticizer*

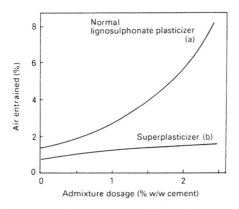

Figure 9.14 *A comparison of the effects on air content of the addition of: (a) a normal lignosulphonate plasticizer, and (b) a superplasticizer*

flowing (self-levelling) concrete which is of particular benefit in floor slabs. Such concrete should set normally and be satisfactory for power floating. Information concerning dosage levels of superplasticizer is shown in *Figure 9.16* whilst workability losses with time after mixing are indicated in *Figure 9.17*. The British Standard relevant to superplasticizers is BS 5075: Part 3: 1985 whose performance requirements are shown in *Table 9.24*.

A comparison between the performance of concretes made using normal plasticizers, superplasticizers and no plasticizers is shown in *Figure 9.18* in respect of compressive strengths. For comparison with American practice the reader is referred to

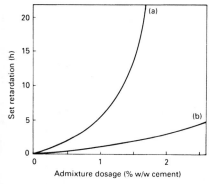

Figure 9.15 *A comparison of the effects on set retardation of the addition of: (a) a normal lignosulphonate plasticizer, and (b) a superplasticizer*

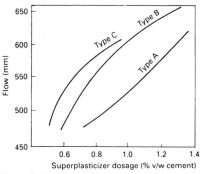

Figure 9.16 *The relative dosages of various categories of superplasticizer needed to produce flowing concretes*

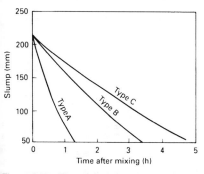

Figure 9.17 *The relative rates of workability loss with various categories of superplasticizer*

Table 9.24 The performance requirements of BS 5075: Part 3: 1985

Property	Test method	Superplasticizing	Retarding superplasticizing
Test-mix A[a]			
Flow	Flow table to BS 1881: Part 105	510–620 mm	510–620 mm
Workability loss on standing, back to that of initial control slump	Slump to BS 1881: Part 102	At 45 min: not less than initial control	At 4 h: not less than initial control
		At 4 h: not more than initial control	
% compressive strength on control	BS 1881: Part 116		
at 7 days		Not less than 90%	Not less than 90%
at 28 days		Not less than 90%	Not less than 90%
Test-mix B[b]			
Slump relative to control	BS 1881: Part 102	Not more than 15 mm less	Not more than 15 mm less
Stiffening time relative to control	BS 4551		
for 0.5 N mm^{-2}	5075 test method	Within 1 h	1–4 h longer
for 3.5 N mm^{-2}		Within 1 h	
% compressive strength on control	BS 1881: Part 116		
at 1 day		Not less than 140%	–
at 7 days		Not less than 125%	Not less than 125%
at 28 days		Not less than 115%	Note less than 115%

[a]Contains plasticizer and has the same water content as the control, giving a high workability flowing concrete.
[b]Contains superplasticizer and has the water content reduced to give equal workability to the control mix.

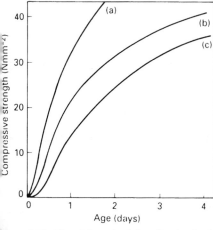

Figure 9.18 *The relative rate of strength gain of water-reduced concretes: (a) superplasticizer; (b) normal water reducer; (c) control*

Table 9.25 which summarizes the performance requirements of ASTM C494.

Retarders It is possible to delay the set of fresh concrete by using a retarding admixture. Although the process is not fully understood it is thought that retardation is caused by the absorption of organic compounds onto the surface of cement compounds thus preventing attack by water. Retarders are dealt with in BS 5075: Part 1: 1982 and ASTM C494, and are classified by retarding power in *Table 9.26*.

Accelerators These are admixtures which accelerate the setting of a mortar or concrete or their hardening processes. As with retarders the mechanism of acceleration is not fully understood. One of the most popular accelerators, calcium chloride, has frequently been misused leading to rapid corrosion of steel reinforcement in reinforced concrete. Up until some years ago the material was mainly available in flake form. This caused non-uniform distribution of calcium chloride which in turn produced excessive galvanic action and corrosion of steel. In more recent times the material has been available in liquid form leading to more uniform distribution. Some authorities have banned the use of this substance. Where it is used it should be strictly controlled to limits indicated in BS 8110: 1986. The use of accelerators is particularly beneficial when concreting at low temperatures. These effects can be clearly seen in *Figure 9.19*.

Sodium chloride is also an effective accelerator and seawater has been used in the casting of unreinforced marine structures.

Table 9.25 The performance requirements of ASTM C494

	Type A, water reducing	Type B, retarding	Type C, accelerating	Type D, water reducing and retarding	Type E, water reducing and accelerating	Type F, water reducing, high range	Type G, water reducing, high range and retarding
Water content (max. % of control)	95	–	–	95	95	88	88
Time of setting, allowable deviation from control (h:min)							
Initial: at least	–	1:00 later	1:00 earlier	1:00 later	1:00 earlier	–	1:00 later
not more than	1:00 earlier, nor 1:30 later	3:30 later	3:30 earlier	3:30 later	3:30 earlier	1:00 earlier, nor 1:30 later	3:30 later
Final: at least	–	–	1:00 earlier	–	1:00 earlier	–	–
not more than	1:00 earlier, nor 1:30 later	3:30 later	–	3:30 later	–	1:00 earlier, nor 1:30 later	3:30 later
Compressive strength (min. % of control)							
1 day	–	–	–	–	–	140	125
3 days	110	90	125	110	125	125	125
7 days	110	90	100	110	110	115	115
28 days	110	90	100	110	110	110	110
6 months	100	90	90	100	100	100	100
1 year	100	90	90	100	100	100	100
Flexural strength (min. % control)							
3 days	100	90	110	100	110	110	110
7 days	100	90	100	100	100	100	100

28 days Length change, max. shrinkage (alternative requirements)	100	90	90	100	100	100	100
Percentage of control	135	135	135	135	135	135	135
Increase over control	0.010	0.010	0.010	0.010	0.010	0.010	0.010
Relative durability factor (min)	80	80	80	80	80	80	80

Table 9.26 Classification of retarders according to retarding power[a]

Group	Slight or negligible retarding effect	Feeble retarding effect	Powerful retarding effect
A	Methanol, sodium formate, formaldehyde, diethoxymethane		
B	Ethanol, calcium acetate, acetaldehyde glycol, glyoxal, oxalic acid, dioxane		Glycolaldehyde, glycolic acid
C	Propyl alcohol, i-propyl alcohol, 1,3-propanediol, 1,2-propanediol, alkyl alcohol, propanol, acetone, propionic acid, acrylic acid 2-chloropropionic acid, malonic acid, 2-hydroxypropionic acid, 1-hydroxypropionic acid (lactic acid)	Glycerin, tartronic acid, butyl acetate, glyceric acid	Acetal, pyruvic acid, glyceroraldehyde, dihydroxy acetone, ketomalonic acid
D	Fumaric acid, aldol Succinic acid anhydride	Maleic acid, erythritol Succinic acid, acetoin	Malic acid, maleic acid anhydride Tartaric acid, dihydroxy tartaric acid, ethyl acetoacetate
E	Chloral hydrate, glycine, EDTA	Diacetone alcohol, acetyl acetone, phoron	α-Ketoglutaric acid, β-ketoglutaric acid, gluconic acid

	urea, adipic acid, 4-hydroxypentan-2-one	8-Hydroxyquinoline	Citric acid, 3% EDTA, sucrose, glucose, fructose, sorbitol, pentaerythritol
	Cupferron (ammonium salt of N-nitrosophenyl hydroxylamine)		
F	Anthraquinone, phenol	Hydroquinone, salicylaldehyde, phloroglucinol, resorcinol, 1,4-naphthoquinone chromotropic acid	Quinone, catechol, pyrogallol, bile acid, 1,2-naphthoquinone, sulphonic acid
G	Zinc oxide (in 3,6-N NaOH solution), beryllium oxide (in 3,6-N NaOH solution), arsenic trioxide, antimony pentoxide	Lead sulphate, lead nitrate, orthoboric acid, antimony trioxide, cadmium oxide, vanadium pentoxide	Zinc oxide
	Red lead, chromium chloride, potassium dichromate, cuprous oxide, cupric oxide, mercury nitrate, mercury chloride, stannous chloride, sodium hypophosphite, potassium pyrophosphate	Cupric nitrate, sodium bisulphite, potassium tetrathionate, sodium tetrathionate, sodium hexametaphosphate, concentrated ammonia	Zinc chloride
			Zinc carbonate, zinc oxide in ammoniacal solution, beryllium sulphate, lead oxide, boron oxide, arsenic pentoxide, sodium pyrovanadate, metaphosphoric acid
			Borax

[a]Unless otherwise stated, all substances were used in an amount of 1% of the cement.

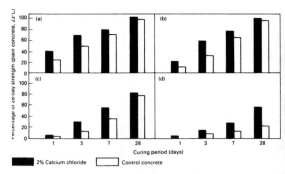

Figure 9.19 *The effects of calcium chloride on strength development at different temperatures: (a) 23°C; (b) 13°C; (c) 5°C; (d) −5°C*

Other accelerators such as calcium nitrite, sodium nitrite, sodium thiocyanate and sodium thiosulphate are used as accelerators and are dealt with in specialist literature. Accelerators are dealt with in BS 5075: Part 1: 1982 and ASTM C494. Performance requirements are shown in *Tables 9.23* and *9.25*.

Corrosion inhibitors The best way to inhibit corrosion of steel reinforcement in reinforced concrete is to surround it by a sufficient quantity of high grade low-permeability concrete. However, particularly in the harsh climates of perhaps Canada, the USA and the former USSR it may be advisable to assist this process with the correct use of chemical inhibitors such as calcium nitrite.

Permeability reducers Concrete by its nature may be a permeable porous material. Permeability arises from the presence of pores as shown by the simplified model (*Figure 9.20*) attributed to T.C. Powers. Even in the ideal but unachievable state of 100% hydration concrete will be about 15% porous. There is a well-defined relationship between percentage hydration and water:cement ratio in order to produce discontinuous pores. This is shown in *Figure 9.21*. Thus any action that can be taken to safely reduce the water : cement ratio will assist in reducing the permeability of the concrete. This can be achieved by the use of permeability-reducing agents such as stearic acids, calcium stearates, oleic acids or isopropyloleate. All such materials should be used with great care; the use of trial mixes is thoroughly recommended.

Reinforcement Concrete is weak in tension and therefore needs to be reinforced where tension occurs. This reinforcement usually takes one of the following forms:

Steel bars — These may be of mild, high tensile or stainless steel which may be coated, for example, with

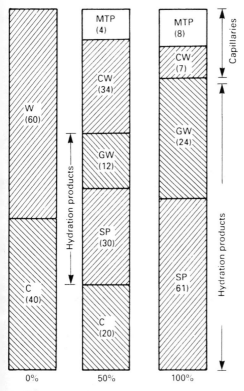

Figure 9.20 *The hydration of a cement paste with a water : cement ratio of 0.473. C, cement; SP, solid product; GW, gel water; CW, capillary water; MTP, empty pores*

epoxy. They may be plain, round or rectangular in cross-section or deformed to increase bond with the concrete.

Steel wires — These may be single-strand or twisted multi-strand. They may be coated and will typically be high-tensile.

Steel fabric — Usually high-tensile steel wire or small cross-section bar laid in an orthogonal pattern and welded at the wire/bar intersections. Fabric may also take the form of small-gauge wire/small mesh galvanized material sometimes used to reinforce cover concrete. This is colloquially referred to as 'chicken wire'.

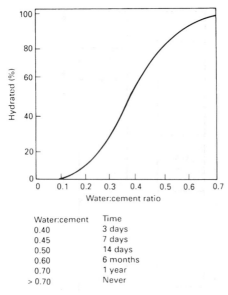

Figure 9.21   *The hydration required to achieve discontinuous pores*

Water:cement	Time
0.40	3 days
0.45	7 days
0.50	14 days
0.60	6 months
0.70	1 year
> 0.70	Never

Steel fibres — Short length small-diameter wire randomly arranged in the concrete.

Glass fibres — As steel fibres.

For steel reinforcement see Chapter 4. For glass fibre reinforcement see Chapter 12.

Other forms of reinforcement, such as bamboo rods, have been used but these are relatively weak and do not form a serious competitor to the above. In special applications such as heavy machine bases it has been customary to use structural steel sections or even rail track sections. Reinforcement may be unstressed at the time of casting or may be prestressed or post-tensioned.

Most reinforced concrete is reinforced by ferrous metal which relies for its longevity on being contained within a sympathetic host material. Corrosion of reinforcement will not occur if the pore solutions of the concrete in which the steel is embedded remain sufficiently alkaline. Under these conditions the formation of a passive film on the metal surface prevents corrosion. However, the carbonation of concrete (the passage of carbon dioxide through the concrete) lowers its alkalinity to a point where the steel becomes de-passivated. If sufficient carbonation occurs and the carbonation front reaches the steel surface the metal will be destroyed. In good quality concrete of low permeability the advance of carbonation is slow and provided sufficient cement has been used its integrity will remain indefinitely.

To be effective, reinforcement needs to be adequate in volume, of the correct type and properly placed. This is a complex subject for which further guidance is given in BS 8110: 1986 and the Institution of Structural Engineers report *Standard Method of Detailing Structural Concrete: 1989* published jointly with the Concrete Society. It is important that the design and detailing of reinforced concrete is entrusted to chartered civil/structural engineers supported by experienced technicians.

9.3.3.8 Mortars, renders and screeds   This section deals with masonry mortars, rendering and floor screeds. These are usually based on cement, lime and sand but may be modified by admixtures and/or polymers.

Mortar   The principal use of mortar is to bond masonry units to produce a continuous load-bearing element. For good durability the mortar must be made from sound materials and compatible with the masonry. There are two critical stages in the life of a mortar—firstly plastic to provide sufficient workability, secondly to provide resistance to the environment and contribute to the strength of the masonry.

Mortars usually comprise sand with one or more of the following: Ordinary Portland cement; blast furnace cement; sulphate-resisting cement; masonry cement; lime; plasticizer; air entraining agents or other admixtures including pigments and water. Recent practice includes the modification of mortars with organic resins. For further information on admixtures see Section 9.3.3.7.

Sands should comply with BS 1200: 1989 and should be hard, clean and free from deleterious materials. Fine sands should only be used if they have a proven record; although they improve workability their use may lead to lower ultimate strength and undue dying shrinkage. Conversely a relatively coarse sand may produce unacceptably low workability.

Cements should comply with BS 12: 1989; BS 146: 1973; BS 4027: 1980 or BS 5224: 1976—the latter contains a proportion of non-cementitious material.

Lime is usually non-hydraulic and can be site-mixed from dry hydrated lime or is available ready-mixed (known as 'course stuff').

The selection of a mortar is often a compromise between many desirable but conflicting properties. *Table 9.27* gives selection guidance.

Mortar performance is closely related to the proportion of binder and in particular cement content. Therefore weigh batching rather than volume batching should be used wherever possible. Ready-to-use retarded cement mortars reduce problems associated with site-mixing operations, particularly inaccurate batching and colour variation. Such mortars are covered

Table 9.27 Mortar mixes[a]

Mortar designation		Type of mortar		
		Cement : lime : sand (by volume)	Masonry cement : sand (by volume)	Cement : sand with plasticizer (by volume)
	i	1 : 0 to 0.25:3	1 : 2.5 to 3.5	1 : 3 to 4
	ii	1 : 0.25:4 to 4.5	1 : 4 to 5	1 : 5 to 6
	iii	1 : 1 to 6	1 : 5.5 to 6.5	1 : 7 to 8
	iv	1 : 2 : 8 to 9	1 : 6.5 to 7	1 : 8
	v	1 : 3 : 10 to 12		

Increasing ability to accommodate movements due to temperature and moisture changes

Increasing strength and improving durability

Increasing resistance to frost attack during construction →

Improvement in adhesion and consequent resistance to rain penetration ←

← (increasing strength and improving durability)

[a]Part of table by courtesy of British Standards Institution. Direction of change in properties is shown by the arrows.

by BS 4721: 1986. Factory-produced ready-to-use mortars consistently lead to higher strengths than site–produced mortars. Where strength is important this is checked by crushing cubes of mortar formed in steel moulds. This is not an altogether simple operation often requiring the mortar to be in the mould for more than seven days to avoid collapse of the cube on de-moulding. The expected strength range of OPC mortars is shown in *Table 9.28.*

It is difficult to assess the strength of in-situ mortar although a specialist analyst may give a guide based on a determination of the mix constituents and proportions.

Winter working of mortars must be restricted to temperatures above 3°C. Because more than 80% of the mass of masonry is in the bricks or blocks the practice of heating aggregates and using steam in place of mixing water is not recommended.

Failures of mortars in service may be both physical and chemical. In general the durability of mortar will increase as cement content increases. Cracking of mortar may be associated with structural, thermal or moisture movement. The addition of lime to a mortar may be beneficial in promoting autogenous healing of hair-line cracks. Mortars can be chemically attacked by sulphates from ground water, some fired-clay products, atmospheric pollution or by products from industrial processes (e.g. acids) or solid-fuel-burning fires. Sulphate-resisting cement should be used in conjunction with bricks of high soluble sulphate content and also to minimize damage from sulphates in groundwater. Fresh mortars can be subjected to frost attack; this can be reduced by the use of air entrainment. Calcium chloride is not recommended as an accelerator because it may result in the mortar being continually damp; it may also promote corrosion in metal wall ties and brick/block reinforcement.

Renders Cement-based coatings can be applied, generally to masonry backings as internal plastering or external rendering to receive applied finishes. External renders will also improve the weather resistance of masonry. The make-up of renders is similar to that of mortar.

Sands should preferably comply with BS 1199: 1976. Where two-coat renders are being applied it may be necesary to remove the coarse particles where it is intended to tool the final coat. Cements recommended for mortars are suitable for renders. Admixtures can be used to retain water and assist workability; for specialist applications polymers and glass fibre can be introduced to reduce shrinkage. Pigments should accord with BS 1014: 1986; depending on the sand used, with some dark colours, some lime bloom may be seen.

External rendering is covered by BS 5262: 1978; internal plastering by BS 5492: 1977. A guide to appropriate mixes is given in *Table 9.29* and *9.30.* The use of premixed or prebagged materials will assist in producing accurately proportioned mixes.

Key points in producing satisfactory rendering include:

Table 9.28 Expected strength range of ordinary Portland cement mortars

Masonry and rendering mortars BS designation	Volume mix proportions		Compressive strength at 7 days ($N\,mm^{-2}$)					
	Cement : sand	Cement : lime sand	Category A		Category B		Category C	
			Satisfactory	High	Low	High	Low	
iii	1 : 6	1 : 1 : 6	0.7–3.0	3.0–7.0	–	>7.0	<0.7	
iii	–	2 : 1 : 9	1.7–4.8	4.8–7.0	0.7–1.7	>7.0	<0.7	
ii	1 : 4.5	–	1.9–6.0	6.0–8.6	0.9–1.9	>8.6	<0.9	
i	1 : 3	1 : 0.25 : 3	4.5–10.4	10.4–12.0	3.0–4.5	>12.0	<3.0	

[a]Category A, satisfactory batching control; category B, batching should be checked; category C, likely to cause failure.

Table 9.29 Mixes suitable for rendering[a]

Mix type	Cement : lime:sand	Cement : ready-mixed lime : sand		Cement : sand (using plasticizer)	Masonry cement : sand
		Ready-mixed lime : sand	Cement : ready-mixed material		
I	1 : 0.25 : 3	1 : 12	1 : 3	–	–
II	1 : 0.5 : 4 to 4.5	1 : 8 to 9	1 : 4 to 4.5	1 : 3 to 4	1 : 2.5 to 3.5
III	1 : 1 : 5 to 6	1 : 6	1 : 5 to 6	1 : 5 to 6	1 : 4 to 5
IV	1 : 2 : 8 to 9	1 : 4.5	1 : 8 to 9	1 : 7 to 8	1 : 5.5 to 6.5

[a]Part of table by courtesy of British Standards Institution.

Table 9.30 Recommended mixes for external renderings in relation to background materials, exposure conditions and finish required[a]

Background material	Type of finish	First and subsequent undercoats			Final cost		
		Severe	Moderate	Sheltered	Severe	Moderate	Sheltered
Dense, strong, smooth	Wood float	II or III	II or III	II or III	III	III or IV	III or IV
	Scraped or textured	II or III	II or III	II or III	III	III or IV	III or IV
	Roughcast	I or II	I or II	I or II	II	II	II
	Dry dash	I or II	I or II	I or II	II	II	II
Moderately strong, porous	Wood float	II or III	III or IV	III or IV	III	III or IV	III or IV
	Scraped or textured	II	III or IV	III or IV	III	III or IV	III or IV
	Roughcast	III	II	II	As undercoats		
	Dry dash	II	II	II			
Moderately weak, porous	Wood float	III	III or IV	III or IV	As undercoats		
	Scraped or textured	III	III or IV	III or IV			
	Roughcast	III	III	III			
	Dry dash						
Metal lathing	Wood float	I, II or III	I, II or III	I, II or III	II or III	II or III	II or III
	Scraped or textured	I, II or III	I, II or III	I, II or III	III	III	III
	Roughcast	I or II	I or II	I or II	II	II	II
	Dry dash	I or II	I or II	I or II	II	II	II

[a]Table partly by courtesy of British Standards Institution.

- proper material specification
- appropriate workmanship (BS 8000: Part 10: 1992) compatibility with background
- attention to detailing at panel edges to prevent water ingress
- care taken not to render over movement joints
- use of light coloured renders to limit thermal movement
- where necessary render should be isolated from background with well-supported lathing
- careful use of metal float finish which may bring laitence to surface and promote crazing
- compatibility of mortar strengths in multi-coat work to avoid shearing between layers.

Screeds By means of a mortar layer screeds provide a level surface to receive subsequent floor coverings or a means of providing falls in flat concrete roofs. They are not intended as a wearing surface. Screed mortars usually comprise cement and sand each of which can be modified or replaced by other materials to enhance performance. Fine aggregate should comply with BS 882: 1983 grading limit M of Table 5 with less than 10% sand passing a $150\mu m$ sieve. Cements should be as for brickwork mortars. Admixtures should conform to BS 5075. Adhesion to sub-base can be improved by using PVA or SBR additives or acrylic polymers.

Ready-to-use screeds are available to BS 4721: 1986. Quick-drying screeds are catered for by proprietary systems to be used strictly in accordance with manufacturer's instructions.

Screeds should be one of the following types:

(1) Monolithic with sub-base; applied within 3 h of casting sub-base.
(2) Bonded to sub-base which should be allowed to harden and be prepared for screed. Screed to be between 25 and 40 mm thick—thicker screeds may curl and crack.
(3) Unbonded and laid on membrane which may act as a dpc. Minimum screed thickness, 50 mm.
(4) Floating, laid on insulating material which should be checked for compressibility. Minimum screed thickness 65 mm.

The water : cement ratio of screeds should be a minimum to obtain adequate workability. Cement : aggregate ratios will normally be in the range of 1:3 and 1:4.5; pan-type mixers will usually produce the best results.

Screeds should be levelled between accurately set battens or strips of prelaid screed. For bonded screeds surface laitence of base concrete should be removed, the base wetted and treated with cement grout and the screed applied immediately. Screeds may be hand or mechanically tamped; where thicker than 50 mm they should be applied in two layers. Curing should be for 7 days followed by at least 7 days of drying out. For sheet or non-ceramic finishes the RH of air immediately above the screed should not exceed 75%. Adequacy of finished screeds may be tested in accordance with BS 8203: 1987.

Defects include surface break-up, curling and hollowness, usually attributable to poor specification, treatment of sub-base, workmanship, curing or selection of thickness.

9.3.3.9 Glass-fibre reinforced cement (GRC)

Introduction Glass-fibre reinforced cement (GRC) is a combination of alkali-resistant glass fibres and a cement/sand mortar. The resultant composite is a concrete-like material combining the compressive properties of cement mortars with the flexural and tensile strength of the glass fibres.

Early attempts to use commercially available silicate glass fibres to reinforce Portland cement failed because of the vulnerability of the fine glass fibres to the highly alkaline environment of the cement matrix. Cemfil alkali-resistant glass fibre was developed by Pilkington following a breakthrough by the UK Building Research Establishment. Pilkington took up the commercial development of this innovation under licence from the National Research Development Corporation and the present generation of GRC products and applications is the result of many years of collaborative development by Pilkington, the BRE and the NRDC, together with the innovative ideas of GRC specifiers and users.

During the rapid expansion of the GRC industry worldwide, the Glass-fibre Reinforced Cement Association (GRCA) was formed in 1975 and has been instrumental in the further development of procedures, codes of practice and standards for GRC materials and products. GRC has wide-ranging applications in the architectural, building, civil and general engineering industries. In typical section thicknesses of 6–20 mm, GRC is widely used as an alternative material to precast concrete, sheet metal, cast iron, timber, plastics and asbestos cement, where the inherent advantages and manufacturing flexibility of GRC, combined with its light weight, non-combustibility, fire resistance and general toughness, make it an appropriate and economical material for use in a range of product forms.

GRC can be composed of different formulations according to the properties required of the finished product. Mostly, however, GRC has a fibre content of 3.5% or 5% by weight (2.9% or 4.1% by volume) combined with a cement/sand mortar. Normal concrete admixtures are commonly used, particularly plasticizers and superplasticizers which allow reduction in the water content. The inclusion of acrylic polymer dispersions in GRC formulations is becoming increasingly popular as it assists curing and improves the toughness and durability. Other admixtures have been used in GRC where there is a need to exploit other properties (e.g. the use of PFA 'cenospheres', perlite or air-entraining agents to obtain a reduction in density or improved fire resistance).

Methods of manufacture vary and include vibration casting, extrusion, injection moulding, spraying and rendering. Each technique imparts different characteristics to the end product.

Clearly then, GRC is not one material with fixed properties, but is a composite with a wide range of possible formulations and resulting properties.

Raw materials The main constituents of glass-fibre reinforced cement (GRC) are cement, sand and alkali-resistant glass fibre. Ordinary and rapid-hardening Portland are the most commonly used cements, although others, including sulphate resisting, pozzolanic and high alumina cement, may also be used. The aggregate will usually be a fine sand with a particle size of 150μm to 1.2 mm, and will be used at aggregate : cement ratios of 0.5–1. Other fine inorganic aggregates and fillers can also be incorporated where particular properties or surface finishes are required.

The water : cement ratio is typically within the range 0.3–0.4. It is often kept to a minimum by the use of water-reducing admixtures. Standard concrete admixtures or those especially formulated for GRC may be used, as appropriate. In some cases, acrylic polymers are used to enhance certain properties.

The glass fibre is supplied either as a continuous roving (which is cut during the GRC manufacturing process into strands of 12–38 mm length) or as precut chopped strands 3–25 mm in length. It is included in the GRC mix at a nominal content of 5% (by weight) for sprayed GRC and at 3.5% (by weight) for pre-mixed GRC.

Details of basic materials such as cement, sand and admixtures are shown elsewhere in this book. The strength of a mortar matrix is beneficially affected by the inclusion of glass fibres as is the resistance to impact. GRC is susceptible to ageing and, if the material remains moist, alkalis from the cement can attack the glass fibre and may produce etching of the surface. It is claimed that these effects are not a problem if proper practice is adhered to. Further research seeks to produce improved formulations of lower alkalinity. It must be stressed that GRC is a complex material for which specialist advice must be sought. For general guidance on properties and usage attention is drawn to *Tables 9.31* to *9.34* and *Figure 9.22*.

9.3.3.10 Sprayed and sprayed-fibre concrete Sprayed concrete can be defined as mortar or concrete conveyed through a hose and pneumatically projected into place from a nozzle. This is sometimes referred to as gunite or shotcrete. In particular applications (e.g. repair of concrete structures, casting of thin shells) there are advantages in using this technique. Sprayed concrete can be produced using a dry or wet mix process (see *Table 9.35* for comparison). The dry process is generally used in the UK. Such concrete may be plain, mesh reinforced, fibre reinforced or normally reinforced for primary

Table 9.31 *The general effects of ageing on the various properties of GRC composites*

Property	Initial stress–strain curve[a]	
	Type A	Type B
Bend-over point	Little change	Little change
Ultimate tensile strength	Decreases to a stable level	Little change
Compressive strength	Little change	Little change
Limit of proportionality	Little change	Little change
Modulus of rupture	Decreases to a stable level	Little change
Strain to failure	Decreases substantially	Little change
In-plane shear strength	Little change	Little change
Interlaminar shear strength	Decreases to a stable level	Little change
Impact strength	Decreases substantially	Decreases
Modulus	Little change	Little change
Poisson's ratio	Little change	Little change

[a]Standard GRC aged in wet conditions

Table 9.32 *Typical initial mean property values of GRC*

Property	Sprayed GRC	Vibration cast (premix) GRC
Dry density ($t\,m^{-3}$)	1.9–2.1	1.9–2.0
Compressive strength (MPa)	50–80	40–60
Young's modulus (GPa)	10–20	13–18
Impact strength ($kJ\,m^{-2}$)	10–25	8–14
Poisson's ratio	0.24	0.24
Bending		
Limit of proportionality (MPa)	7–11	5–8
Modulus of rupture (MPa)	19–31	10–14
Direct tension		
Bend-over point (MPa)	5–7	4–6
Ultimate tensile strength (MPa)	8–11	4–7
Strain to failure (%)	0.6–1.2	0.1–0.2
Shear		
In-plane (MPa)	8–11	4–7
Interlaminar (MPa)	3–5	NA[a]

[a]NA, not applicable.

structural members. Fibre reinforcement significantly alters the material behaviour by improving post-crack ductility.

In the dry-mix process, cement, dry or moist aggregate and other dry admixtures are batched/mixed then fed into a pur-pose-made gun. The mix is pressurized, metered into a compressed air stream and conveyed through a delivery hose to a nozzle where water is introduced under pressure as a spray to wet the mix before it is projected into place.

Table 9.33 *Estimated long-term stable property values for standard GRC*

Property	Sprayed GRC	Vibration cast (premix) GRC
Compressive strength (MPa)	75	50
Modulus (GPa)	25	20
Impact strength (kJ m^{-2})	4	3
Poisson's ratio	0.24	0.24
Limit of proportionality (MPa)	10	8
Modulus of rupture (MPa)	13	10
Bend-over point (MPa)	5.5	4.5
Ultimate tensile strength (MPa)	5.5	4.5
Strain-to-failure (%)	0.04	0.03
In-plane shear strength (MPa)	5.5	4.5
Interlaminar shear strength (MPa)	4	NA

Table 9.34 *Typical design stresses[a] for the GRC formulations indicated in Table 9.32*

Stress type	Loading example	Design stress (MPa)	
		5% spray GRC	3.5% premix GRC
Compressive	Compressive	12	12
Bending	Bending solid beams or plates	6	4
Tensile	Cylindrical hoop stresses	3	2
Tensile	Bending sandwich panels	3	2
Web shear	In-plane shear of webs in box sections	2	1
Bearing shear	Shear loading at bearing positions	1	1

[a]These design stresses were obtained by applying a reduction factor of approximately 1.8 to the long-term cracking strength (e.g. limit of proportionality and bend-over point) of the GRC. The values apply to cases where the characteristic strength values are not less than: modulus of rupture—sprayed GRC 19 MPa, premix GRC 8 MPa; limit of proportionality—sprayed GRC 6.5 MPa, premix GRC 5 MPa.

In the wet-mix process cement, aggregate, admixtures and water are batched/mixed before being fed into delivery equipment. The mix is metered into a delivery hose and conveyed to the nozzle where compressed air is used to project material into place.

These techniques suffer from material rebound. Factors involved include direction of spraying, spraying distance, aggregate grading and moisture content, water pressure, nozzle design, admixtures, layer thickness and surface conditions.

Table 9.35   *Comparison of wet and dry processes*

Characteristic	Dry process	Wet process
Water content	Variable, controlled by nozzleman	Constant, controlled during batching
Use of admixtures	Accelerator occasionally used	Plasticizers and accelerators common
Stop/start flexibility	Good	Poor, concrete can set in hoses
Material rebound	High, 15–35% on vertical surface	Low, 10–15% on vertical surface
Homogeneity of concrete	Fluctuating water:cement ratio causes variation	Pumping can cause segregation
In situ properties	Good compaction and strength	Less remarkable but less variable
Spraying environment	Usually dusty	Relatively clean

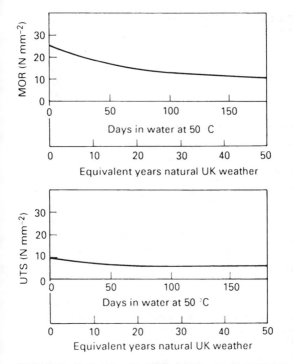

Figure 9.22   *Typical strength variation with time for standard hand-sprayed GRC containing 5% alkali-resistant fibre. Sand : cement ratio 0.5; water : cement ratio 0.3. UTS, ultimate tensile strength; MOR, modulus of rupture*

The relationship between layer thickness and % rebound is shown in *Figure 9.23*. The use of silica fume is understood to reduce rebound; the effect of this is illustrated in *Figure 9.24*.

All materials used with this technique are dealt with elsewhere in this book.

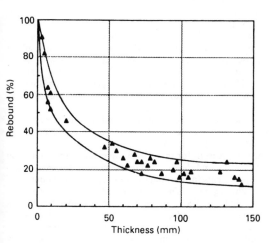

Figure 9.23 *Effect of layer thickness on material rebound (Parker et al., 1977)*

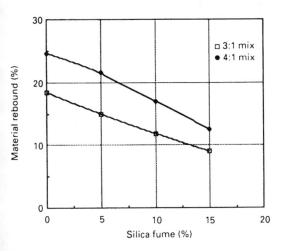

Figure 9.24 *Effect of silica fume on material rebound*

9.3.3.11 Concrete blocks Examples of the use of concrete date as far back as 5600 BC, but the concrete block, as we know it today, based on the use of Portland cement, was not introduced into the UK until around 1850. Joseph Gibb patented a process to manufacture a product to imitate the dressed stone of that period. The blocks so produced were generally hollow with moulded faces.

It was not until the early 1900s, coinciding with the significant growth in the production of Portland cement, that the concrete-block industry became established. The first noticeable growth between 1918 and 1939 came about as a result of the house-building programme following World War I. These were mainly clinker blocks and used for partition walls in houses. However, the most significant period of development and growth did not occur until the building programme following World War II. Since that time, the growth of concrete blocks has, indeed, been notable and has increased from around 4 million m^2 in 1955 to around 110 million m^2 in 1988, as shown in *Figure 9.25*.

The main reasons for this growth were due to the promotion of cavity walls, and the steady improvements in thermal-insulation requirements for dwellings resulting in the development of lightweight aggregate blocks, and autoclaved aerated concrete blocks. That, coupled with their low cost, lighter weight and ease of handling, provided economy in terms of time and cost of construction. Further developments have taken place which demonstrate means of improving the efficiency of masonry construction without sacrificing either the important aspect of freedom of design or the potential for achieving very high standards of quality and performance.

Developments have taken place with the product and, in addition to the common and insulating blocks, a wide range

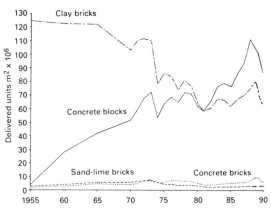

Figure 9.25 *Deliveries of bricks and blocks in the UK since 1955*

of high-quality facing units have emerged. Realization of their structural potential has resulted in the development of the strength range. The increase in the use of concrete blocks and concrete bricks is such that concrete masonry is currently the major masonry material used in the UK.

Types of block Concrete masonry units (blocks) are covered by BS 6073: Parts 1 and 2: 1981. Types of block include:

(1) *Solid block*. These contain no formed holes other than those inherent in the material. Autoclaved aerated blocks are thus by definition 'solid'.
(2) *Cellular block*. These have one or more formed holes which do not wholly pass through the block. Voids are generally used to lighten block and improve insulation. They are laid closed end uppermost.
(3) *Hollow block*. Those with one or more cavities passing right through the block. In addition to normal use they may be used for reinforced blockwork.
(4) *Common block*. Those units which are used for general construction work where appearance is not of paramount importance (e.g. below ground).
(5) *Facing block*. Specially made or selected to have consistent shape/texture.
(6) *Architectural masonry block*. Often used to describe a higher than normal specification facing block. May have special finish such as exposed aggregate.
(7) *Insulating block*. Usually manufactured from lightweight aggregate or autoclaved aerated concrete to give better than normal inherent insulating properties. May also have insulating materials incorporated within the block and be referred to as composite insulated blocks.

Properties of blocks BS 6073 lays down a maximum block dimension of 650 mm to avoid confusion with other precast slabs or panels. Blocks are available in a range of sizes, most of which are indicated in *Table 9.36*.

The permitted deviation of block dimensions is shown in *Table 9.37*. The dimensions shown in *Table 9.36* are work sizes (10 mm less than the coordinating space to allow for nominal mortar joints).

The density of concrete blocks ranges from about $475 \, \text{kg m}^{-3}$ for the lightest autoclaved aerated blocks to around $2000 \, \text{kg m}^{-3}$ for dense aggregate blocks. This produces blocks in the weight range of 3 kg to 40 kg at the upper end of which it may be necessary to provide mechanical handling.

The compressive strength of blocks ranges from $2.8 \, \text{N mm}^{-2}$ to $35.0 \, \text{N mm}^{-2}$ (see BS 5628: Part 1: 1978 and Part 2: 1985). Concrete blocks are inherently durable but full recommendations for minimum quality of blocks and mortar are given in BS 5628: Part 3: 1985. The linear movement due to temperature changes in blocks is generally considered to be reversible and

Table 9.36 *Work sizes of blocks*

Length (mm)	Height (mm)	Thickness (mm)														
		60	75	90	100	115	125	140	150	175	190	200	215	220	225	250
390	190	*	*	*	*	*		*	*		*	*				
440	140	*	*	*	*			*	*		*	*	*		*	
440	190	*	*	*	*			*	*		*	*	*	*		
440	215		*	*	*	*	*	*	*	*	*	*	*	*	*	*
440	290	*	*	*	*			*	*		*	*	*			
590	140		*	*	*			*	*		*	*	*			
590	190		*	*	*			*	*		*	*	*			
590	215		*	*	*		*	*	*	*		*	*		*	*

typically in the range of $7-14\times10^{-6}$ per °C. Concrete blocks undergo dimensional changes as a result of moisture variation. Units will expand on wetting and shrink on drying. Since the units contain more water than is necessary to hydrate the cement long-term shrinkage will occur until the blocks dry to equilibrium. Maximum permitted shrinkage limits are shown in *Table 9.38*.

Basic specification data for block materials is covered elsewhere in this book.

Blockwalling Concrete blockwork is a combination of units and mortar. The result may fulfil a variety of functions such as providing walling for housing, industrial, commercial and civic buildings. Blocks may be used in conjunction with clay brickwork in cavity construction, provided caution is exercised to recognize possible differential movements between the dissimilar materials. Concrete masonry diaphragm walls provide considerable strength gains from a small increase in materials. One study indicates an eight-fold increase in lateral strength for an 8% increase in material quantity. The inherent weakness in tension may be overcome in some applications by grouting steel reinforcement into voids in hollow blockwork to produce a composite construction similar to reinforced concrete (see BS 5628: Part 2).

Mortars are extensively dealt with in Section 9.3.3.8 and attention is particularly drawn to *Tables 9.27* and *9.28*.

Properly designed and built concrete masonry walls can give adequate resistance to rain. There are two ways in which walls resist rain penetration:

(1) A wall built using high absorption units may absorb and retain water and then subsequently dry out.

Table 9.37 *Dimensional deviations of blocks*

Dimension	Maximum deviation (mm)
Length	+3 − 5
Height	+3 − 5
Thickness	+2 − 2 average
	+4 − 4 at any individual point

Table 9.38 *Maximum permitted shrinkage of concrete masonry units*

Material	Shrinkage (%)
Concrete bricks and dense and lightweight aggregate block	0.06
Autoclaved aerated concrete blocks	0.09

(2) A wall built of impermeable units and set in full joints of strong cement : lime will resist the rain.

Further guidance is given in BS 5628 Part 3; BS 4315: 1983 indicates test rig information should it be desirable to check validity of design by full-scale testing.

Guidance on the design of all types of blockwalling is given in BS 5628 Parts 1 and 2. When comparing blocks with bricks of given compressive strength it is important to take into account the differing aspect ratios. For example a similar wall can be achieved by using a 215 mm × 100 mm solid block of strength 7 N mm^{-2} as would be achieved using a typical brick of strength 15 N mm^{-2} to 20 N mm^{-2}.

The level of thermal insulation for buildings is laid down in Building Regulations. Energy is becoming an increasingly expensive resource; to meet that challenge thermal transmittance values are being reduced. For dwellings and most other buildings the U value of 0.6 W m^{-2} required in 1985 was reduced to 0.45 W m^{-2} K^{-1} in 1990. Such reductions have led to manufacturers producing blocks of higher thermal efficiency. Trade literature should be consulted for up-to-date information.

Examples of walls able to satisfy the sound insulation requirements of the Building Regulations are given in DOE Approved Document E. There is a relationship between the mass of a wall and its sound insulation, although efficiency is also affected by stiffness, permeability and absorption characteristics. The sound insulation of cavity walls is related not only to surface mass but also to the width of cavity and the rigidity and spacing of wall ties.

The fire resistance of a wall will depend on a number of factors including thickness, block type (hollow or solid) and aggregate type (dense, lightweight or aerated). Typically a 100 mm solid block wall will provide a notional fire resistance of 2 h.

The durability of blockwalling will vary with the nature of the units, the composition of the mortar, the exposure to weather (and other abnormal temperature effects), pollution and aggressive conditions. Full recommendations are given in BS 5628: Part 3.

9.3.3.12 Concrete tiles and slates The increased specification for and refurbishment of pitched roofs has meant a demand for roofing products capable of meeting the needs of modern technological design.

In this section, man-made concrete tiles and slates are described, and reflect their contribution in providing durable, aesthetic, and cost-effective alternatives to natural roofing materials.

As well as the previous factors, planning authorities have strict policies with regard to the choice and colour of roofing materials for use within conservation areas or on listed buildings. Re-roofing with heavier materials can also involve requirements to comply with Building Regulations in respect of structural stability.

Modern concrete tiles have been gradually changing their utilitarian and functional image to one of style, variety and liveliness. Since their inception in Britain over 60 years ago, concrete tiles have been backed up by guarantees of durability. The material is ideally suited to high-speed, computer-controlled production, enabling it to be made widely available in high quality and at low cost. This versatility allows the product to be produced in a wide range of profiles and colours.

Perhaps the most important development has been the introduction of the interlocking slate. The ease of installation and improved performance characteristics will mean that, in some areas, cost savings can be made in installation and in the supporting structure. The use of new materials technology has enabled a thinner and stronger product to be made.

Only a few manufacturers are producing authentic copies of old tile designs and fittings. The ability of manufacturers to supply specialized products is often an important factor to architects wishing to specify sensitively designed products for conservation projects.

Dry-fix systems are becoming more popular because they allow roofs to be completed in all weathers and eliminate the need for traditional mortar at ridges and verges. Dry-fix systems have been developed by the leading concrete-tile manufacturers to suit most product ranges and incorporate ventilation apertures allowing an unrestricted air flow to overcome condensation in insulated roof spaces. Apart from the ease of installation, dry-fix systems mean that frost and wind damage are negligible and the roof becomes virtually maintenance free.

Research into new materials technology will, in the future, play a far greater role in the development of roofing products. However, for as long as today's modern buildings continue to be built using traditional methods and materials, the roof-coverings market will remain dominated by slates and tiles.

Concrete tiles These should comply with BS 473.550: 1990 and be installed to BS 5534: Part 1: 1990 and Part 2: 1986 which gives guidance on fixings to resist wind forces. Tiles are manufactured from sand (see Section 9.3.3.); cement (usually to BS 12: 1978); pigments (inert, colour-stable inorganic oxides; resistant to weathering, atmospheric conditions, alkali attack, degradation by ultra-violet light and sympathetic to the concrete mix); and (potable) water. They are continuously extruded onto profiled moulds and then subjected to pressures in excess of 500 N mm^{-2}. Curing takes place at around 40°C and high humidity for 8–24 h. Coloured polymer emulsions are then sometimes surface-applied to suppress efflorescence. Tiles are classified as follows (see *Table 9.39*):

(1) Double lap plain (see *Figure 9.26(a)*)

Table 9.39 Types of concrete roofing tile

Type of tile	Size (mm)	Max. gauge (mm)	No. of tiles per m²	Metre run of battens per m²	Weight (75 mm lap) (kg m)⁻³	Min. pitch (°)	
						Smooth surface finish	Granuled finish
Plain	267 × 165	100	60	10	78	35	35
Interlocking							
Bold roll	420 × 330	345	9.7	2.9	48	17.5	30
Double pantile	420 × 330	345	9.8	2.9	47	22.5	25
Double Roman	420 × 330	345	9.7	2.9	45	22.5	25
Concrete slate	420 × 330	345	10.0	3.0	52	17.5	–
Single pantile	387 × 230	312	15.9	3.2	48	25	30
Interlocking trough	387 × 230	312	16.0	3.2	49	25	30

(2) Single lap interlocking (see *Figure 9.26(b)*).

The plain tile is a versatile product and suited for use in complex roof shapes where dormers, valleys and hips require a small-scale roofing module. The batten gauge is

$$\frac{(\text{length of tile}) - (\text{length of lap})}{2}$$

Laps should not exceed 90 mm. As shown in *Figure 9.27(a)*, double lap means that the top of any tile is covered by two tiles and at any section over the battens there are three thicknesses of tile. It is also worth noting that the effective pitch of a double-lap plain tile is about 7° less than the rafter pitch (see *Figure 9.27(b)*). The size and format of the concrete plain tile are

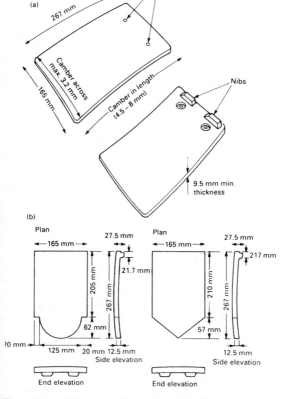

Figure 9.26 *(a) Standard concrete double-lap plain tile-top and underside. (b) Plain feature tiles*

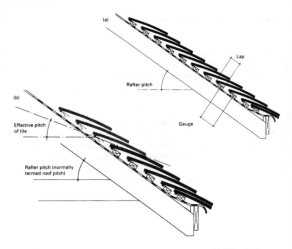

Figure 9.27 *(a) Rafter pitch for double-lap tiles. (b) Tile pitch versus rafter pitch*

similar to its clay counterpart. They are manufactured to a size of 267 mm × 165 mm and have a 'camber' over both length and width. The camber is intended to reduce the capillary action of rainwater which can be drawn both upwards and sidewards.

Interlocking tiles have long been used to reduce the number required to cover a given area. Typical interlocking tile designs are illustrated in *Figure 9.28*.

The number, position and size of the anti-capillary grooves and ribs can greatly improve the resistance to driving rain by providing small cavities which capture wind-driven rain. The side-lock is normally designed with a minimum of gaps but allowing some lateral adjustment of approximately 3 mm with maximum flexibility for optimum water-carrying capacity and resistance to driving rain (see *Figure 9.29*). Tile profile also has a significant effect on resistance to wind loads. The maximum resistance to wind uplift can be achieved by optimizing the ratio of tile weight to tail thickness thereby achieving a centre of gravity closer to the leading edge of the tile. Surface finish and profile are also relevant in minimizing the localized effects of wind. The tile pitch for a single-lap interlocking tile is approximately 5° less than the rafter pitch. The practical minimum for a flat concrete slate tile is 17.5° although some manufacturers offer products which can be used at roof pitches of 15° and 12.5°.

Concrete tiles are produced in a wide range of surface textures (granular or smooth faced), finishes and colours. Surface texture and mature appearance can be achieved by an application of fine silica sand granules. This assists the weathering process but may, in some areas, encourage moss growth.

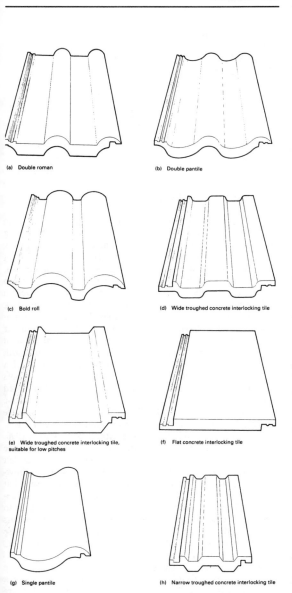

(a) Double roman

(b) Double pantile

(c) Bold roll

(d) Wide troughed concrete interlocking tile

(e) Wide troughed concrete interlocking tile, suitable for low pitches

(f) Flat concrete interlocking tile

(g) Single pantile

(h) Narrow troughed concrete interlocking tile

Figure 9.28 *Various designs of concrete interlocking tiles*

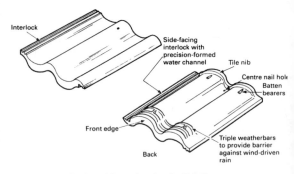

Figure 9.29   *Back and face of an interlocking tile*

There is no British Standard wind/driving rain test although BRE have developed a method of test. Permeability testing should accord with BS 473.550. Concrete tiles having low water absorption (around 7% of dry weight) have good frost resistance. Some have satisfactorily achieved, in Scandinavia, prolonged freeze–thaw cycles with temperatures down to −20°C. Conversely concrete tiles have also performed well in high temperatures (up to 70°C surface temperature on dark coloured tiles). Such tiles also exhibit good resistance to atmospheric pollution having performed well in areas with sulphur dioxide levels up to $70\mu\mathrm{g\,m}^{-3}$. Above this level accelerated surface erosion may occur. Impact resistance is high; in South Africa concrete roof tiles are understood to have resisted a 45 mm diameter hailstone with impact energy of 20 J. Roofs must be capable of resisting anticipated dead and imposed loads. Tiles, where necessary, should be tested for transverse strength in accordance with BS 473.550. It is normal practice to accept the average strength from six samples selected for test.

Tiles should be laid in accordance with BS 5534: Part 1: 1990. Typical roof loadings of tiles are shown in *Table 9.40*. The common practice in England is to lay reinforced bitumen felt type BS 747: 1F: 1977 draped over rafters. In Scotland, however, the practice is to use rigid sarking or boarding. Such felt should comply with BS 3177: 1959 tests in respect of permeability. When constructed it is important that in addition to structural considerations roofs comply with the requirements of the Building Regulations and if applicable National House Building Council (NHBC) in respect of thermal insulation, ventilation, sound insulation and fire resistance.

In a comparatively new departure some manufacturers are producing tiles made using lightweight aggregates which have a laid weight in the order of $28\,\mathrm{kg\,m}^{-2}$. In re-roofing they are therefore comparable to slates for which they may be substituted with minor if any structural upgrading.

Table 9.40 *Typical roof loadings (kg m^{-2}) at 30° rafter pitch*

Roof tile/slate	Slope load				Plan load*		
	Tile	Batten	Underlay	Dead load	Dead load	Snow load (imposed)	Total load
Reconstructed stone slates (475 mm × 457 mm)	80.0	2.56	1.20	83.76	96.72	76.50	173.22
Concrete plain tile	78.20	4.30	1.20	83.70	96.64	76.50	173.14
Concrete interlocking tile	45.50	1.49	1.20	48.19	55.64	76.50	132.14
Lightweight concrete interlocking tile	28.0	1.49	1.20	30.69	35.43	76.50	111.93
Polymer-modified-cement concrete interlocking slates	23.5	1.94	1.20	26.64	30.76	76.50	107.26
Fibre–cement slate (600 mm × 300 mm)	21.5	1.94	1.20	24.64	28.45	76.50	104.95

*Load on plan = [Load on slope (including felt and battens) × cos(roof pitch (at 30° = 0.866))].

Concrete slates   These are produced in four basic categories:

(1) Concrete reconstructed stone slates
(2) Fibre-cement slates
(3) Resin- and polymer-bonded slates
(4) Polymer-modified-cement slates.

Table 9.41 provides basic details of concrete slates whilst *Figures 9.30* and *9.31* show examples of reconstructed stone slates and polymer-modified-cement slates respectively.
   Points of interest concerning these four types are:

(1) *Concrete reconstructed stone slates*
   These are made to resemble natural slate from areas such as the Cotswolds, Derbyshire, Wales and Yorkshire. Most designs are for double-lap laying although interlocking single-lap versions are available. There is no specific British Standard for these slates although most comply with BS 473.550 in respect of strength and durability. The raw materials used are basically sand, cement, oolitic lime stone and water—usually blended with fragments of natural stone.
(2) *Fibre-cement slates*
   These were produced for many years using cement bonded with asbestos fibres. Following confirmation of the health hazard from asbestos, synthetic fibres are now used. Asbestos slates were produced to BS 590: Part 4: 1974 synthetic fibre slates are usually covered by BBA Certification. Materials used are basically OPC to BS 12 synthetic cellulose fibres and water. These slates are available in sizes ranging from 600 mm × 300 mm to 400 mm × 200 mm in a range of finishes from semi-matt to the randomly textured/riven surface of dressed natural slate. Watertightness of assembled slating can be checked by test; permeability can be reduced by surface application of an acrylic which also reduces adhesion of mosses and lichens. Maximum water absorption should not exceed 18%. These slates have densities in the range 1600 kg m^{-3} to 1800 kg m^{-3} when tested in accordance with BS 4624 1981. Fibre-cement slates can all achieve a class AA rating to BS 476: Part 3: 1958 and some achieve class-O as defined by Building Regulation E15.
(3) *Resin- and polymer-bonded slates*
   These are in direct competition with fibre-cement slates and are of comparatively recent manufacture. Raw materials are natural aggregates of crushed stone/slate reinforced with glass fibre in a binder of polyester or acrylic resin. In some versions they are up to 15% lighter than the equivalent fibre-cement slates. A wide colour range is produced but slates may be sensitive to colour loss in ultraviolet light. These slates are highly resistant to air-borne pollution impact and adverse weather. The resinous surface offers

Table 9.41 Types of man-made slates

Type of slate	Size (mm)	Max. gauge (100 mm lap) (mm)	No. of slates per m² (100 mm lap)	Metre run battens per m²	Weight (kg m⁻²)	Min. pitch (°)
Fibre cement (double lap)	600 × 300	250	13.3	4.0	21.0	20
	500 × 250	200	20.0	5.0	19.5	20
	400 × 200	155	31.9	6.45	21.6	20
Reconstructed stone slates						
North England	457 × 305 to 686 × 457	190–279	17.2–7.82	3.58–5.24	80–99	15–25
Cotswold	Length 300–550 Width 200–500	110–200	22 (80 mm lap)	6.25	82	30
Interlocking slate						
Resin bonded	300 × 336	250	13.3 (75 mm lap)	4.0	17	25
PMC concrete*	325–330	250	13.3 (75 mm lap)	4.0	23.5	25

*PMC, polymer-modified cement.

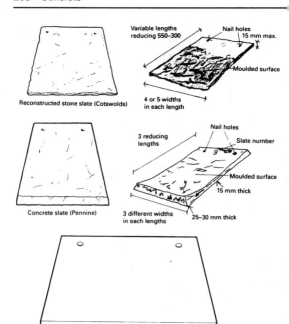

Figure 9.30 *Examples of reconstructed stone slates*

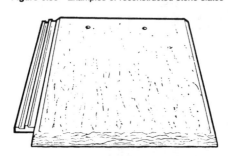

Figure 9.31 *Polymer-modified-concrete interlocking slate*

low adhesion to mosses and lichens. These slates may achieve only a Class 2 fire rating thus inhibiting their use in some circumstances.

(4) *Polymer-modified-cement slates*

Polymer-modified-cement mortar is one where part of the mixing water is replaced by an aqueous polymer emulsion. This produces high workability with low water : cement ratios. The mortar has low void content, high flexural strength and improved chemical and frost resistance.

Crushed stone/slate granules are used as aggregate. Such slates also have low water absorption (< 1%), no tendency to curl once cured, lower surface erosion than normal concrete and high resistance to ultraviolet light. Acrylic surface coating will also reduce formation of efflorescence and adherence of mosses and lichens. Because the polymer content (typically 5%) is relatively low a high fire resistance is achievable with a Class-O rating for spread of flame.

Relative costs of tiling and slating systems Costs of labour, plant and materials vary with time but *Table 9.42* gives a useful comparison between various roof coverings. Accurate costings should be obtained for any specific applications.

9.3.3.13 Other products In addition to those described in earlier sections, concrete is used to produce a wide range of products. These include:

Flags (or paving stones) These are commonly produced in sizes 900 mm × 600 mm and 600 mm × 600 mm to 50 mm nominal thickness. They will normally comply with BS 368: 1971 (AMD 1976, 1979 and 1986) which covers materials, finish and colour, casting and curing, dimensions, test requirements, sampling; test methods for measuring flags, transverse strength, water absorption.

Pipes These are manufactured in a wide range of diameters/ wall thicknesses. Pipes between 300 mm and 600 mm are often unreinforced; those of 675 mm diameter and above are usually steel mesh reinforced. Diameters up to 1800 mm are readily available; larger pipes would usually be purpose-made. Most manufacturers provide a large range of jointing techniques and ancillary equipment such as manholes, inspection chambers, gullies and angled bends to match their standard ranges. *Table 9.43* provides some basic data on readily available sizes. Pipes are available in OPC and SRPC and will normally conform to BS 5911: Part 100: 1988.

Concrete sleepers These are usually prestressed, precast concrete units. They are used as an alternative to traditional timber sleepers beneath rail tracks. Information on these may be obtained from FIP report No. 537: 1987. There is currently no British Standard covering these units.

9.3.3.14 Health and safety Portland and other hydraulic cements are harmless in normal use. However, alkalis are released when water is added and direct contact of freshly mixed concrete with the skin should therefore be avoided. The abrasive nature of the aggregate can aggravate the effects of alkalis on the skin. Any concrete or mortar on the skin should be removed with soap and water. If cement enters the

Table 9.42 *Relative roofing costs (index 100 = £13.50): price index includes materials and labour per m² of laid roof*

Slate/tile type	No.	Cost Material	Cost Labour
Man-made slates 600 mm × 300 mm laid to 100 mm lap on 38 mm × 25 mm softwood battens on BS 747: 1F underfelt	167	£17.60 78%	£4.90 22%
Concrete interlocking tiles 420 mm × 330 mm laid to 75 mm lap on 38 mm × 25 mm softwood battens on BS 747: 1F underfelt	100	£11.50 85%	£2.00 15%
Concrete plain tiles 267 mm × 165 mm laid to 65 mm lap on 38 mm × 25 mm softwood battens on BS 747: 1F underfelt	203	£22.40 82%	£4.90 18%
Reconstructed stone slates 457 mm × 457 mm laid to 76 mm lap on 38 mm × 25 mm softwood battens on BS 747: 1F underfelt	284	£28.40 80%	£7.10 20%
Lightweight concrete interlocking tiles 420 mm × 330 mm laid to 75 mm lap on 38 mm × 25 mm softwood battens on BS 747: 1F underfelt	128	£14.80 85%	£2.60 15%
Polymer-modified-cement concrete interlocking slates 325 mm × 330 mm laid to 75 mm lap on 38 mm × 25 mm softwood battens on BS 747: 1F underfelt	191	£21.40 83%	£4.40 17%
Clay plain tiles 265 mm × 165 mm laid to 65 mm lap on 38 mm × 25 mm softwood battens on BS 747: 1F underfelt	259	£30.20 86%	£4.90 14%

Table 9.43 *Concrete pipes: basic data*

Type	Class	Unreinforced					Reinforced							
		300	375	450	525	600	675	750	900	1050	1200	1350	1500	1800
Full length pipes	L	•	•	•	•	•	•	•	•	•	•	•	•	•
	M	•	•	•	•	•	•	•	•	•	•	•	•	•
	H		•	•	•	•	•	•	•	•	•	•	•	•
Rocker pipes	L	•	•	•	•	•	•	•	•	•	•	•	•	•
	M	•	•	•	•	•	•	•	•	•	•	•	•	•
	H		•	•	•	•	•	•	•	•	•	•	•	•
Spigot/plain end pipes	L	•	•	•	•	•	•	•	•	•	•	•	•	•
	M	•	•	•	•	•	•	•	•	•	•	•	•	•
	H		•	•	•	•	•	•	•	•	•	•	•	•
Socket/plain end pipes	L	•	•	•	•	•	•	•	•	•	•	•	•	•
	M	•	•	•	•	•	•	•	•	•	•	•	•	•
	H		•	•	•	•								
Perforated full length pipes	L	•	•	•	•	•								
	M	•	•	•	•	•								
	H		•	•	•	•								

Notes: L = light; M = medium; H = heavy. (Courtesy: Hepworths Building Products)

eye, it should be washed out with plenty of clean water and medical treatment sought without delay.

Protective clothing should be worn and care taken to avoid getting concrete into wellington boots. Clothes contaminated with cement should be thoroughly cleaned before re-use. The use of waterproof gloves is strongly advised.

References

AFNOR, Granulats—essai au bleu de méthylène, *Experimental Standard*, p. 18–592, AFNOR, Paris (1980)

BRITISH STANDARDS INSTITUTION, *BS 812: Part 123 Draft, Concrete prism method. Testing aggregates. Methods for the assessment of alkali–aggregate reactivity potential*, Milton Keynes (1988)

CANADIAN STANDARDS AUTHORITY, *CSA A23.2–14A: Alkali–aggregate reaction (concrete prism test)*. Methods of Test for Concrete, Ottawa, pp. 183–185 (1977)

CHATTERJI, S., An accelerated method for the detection of alkali–aggregate reactivities of aggregates, *Cement and Concrete Research*, **8**, 647–650 (1978)

HOBBS, D.W., Testing for alkali–silica reactivity, *Cement and Concrete Association Test Methods*, Cement and Concrete Association, Wexham Springs, Slough (1985)

JONES, F.E. and TARLETON, R.D., Recommended test procedures. Part VI: Alkali–aggregate interaction. Experience with some forms of rapid and accelerated tests for alkali–aggregate reactivity, *National Building Studies Research Paper 25*, HMSO, London (1958)

MING-SHU, T., A rapid method for identification of alkali-reactivity of aggregate, *Cement and Concrete Research*, **13**, 417–422 (1983)

OBERHOLSTER, R. and DAVIES, G., An accelerated method for testing the potential reactivity of siliceous aggregates, *Cement and Concrete Research*, **16**, 181–189 (1986)

PARKER, H.W., FERNANDEZ-DELGADO, G. and LORIG, L.J., A practical new approach to rebound losses, *Shotcrete for Ground Support (SP-54)*, American Concrete Institute, Detroit, 149–187 (1977)

STARK, D., Osmotic cell test to identify potential for alkali–aggregate reactivity, *Proceedings of the 6th International Conference on Alkalis in Concrete*, Copenhagen, pp. 351–357 (1983)

TAMURA, H., An experiment on rapid identification of alkali-reactivity of aggregate, *Review of the 38th General Meeting of the Cement Association of Japan*, Tokyo, pp. 100–103 (1984)

Bibliography

DORAN, D.K. (ed.), *Construction Materials Reference Book*, Butterworth-Heinemann, Oxford (1992):
  Chapter 14: Pomeroy, C.D.
  Chapter 15: Hodgkinson, L.
  Chapter 16: Pitts, J.
  Chapter 17: Tovey, A.K.
  Chapter 18: Buttler, F.G.
  Chapter 19: Hodgkinson, L.
  Chapter 20: Cross, S.H. and Ridd, P.J.
  Chapter 21: Sykes, R.A.
  Chapter 22: Hodgkinson, L.
  Chapter 23: Kawamura, M.
  Chapter 24: Robins, P. and Austin, S.
  Chapter 25: Dodd, J.

10 *Cork*

10.1 Introduction

Cork is obtained from the bark of the cork oak tree (*Quercus Super* L.) found primarily at the western end of the Mediterranean in Portugal, Spain and North Africa. The bark thickens and develops with age and can be harvested as a regular crop (every 9–12 years) from commercial plantations and natural forests. Cork bark is composed of cells which are five-sided, impermeable and contain a large proportion of air. This cellular structure allows cork to be compressed and to recover its original dimensions without extrusion—a property yet to be equalled by synthetic products used in this field. Cork is, therefore, a perfect natural insulant.

Because cork is a natural material it regenerates itself without depletion of its source. It has a place in every phase of a building.

10.2 Manufacture

Corkboard is manufactured by baking at a temperature of 350°C and a pressure of 30 000 kg m^{-2}. There is no contact with air during this process and the cork's natural resins are released from the cells causing the granules to bond together, becoming dark brown in colour. After an appropriate period of 'cooking', a block of corkboard is produced which is cooled, rested, trimmed to size and cut into slabs.

This process causes natural cork granules with a density of approximately 200 kg m^{-2} to release volatile elements with a commensurate decrease in density. At the same time expanded cork granules are compressed causing their compaction and agglomeration to form a cork block with a higher density than that of expanded granules. This density can vary from 80 kg m^{-3} to over 600 kg m^{-3} depending on:

(1) Degree of compression
(2) Quality of cork
(3) Granule size
(4) Purity of granulation
(5) Manufacturing process
(6) Temperature.

The combination of these factors enables the manufacture of different types and densities for numerous purposes.

0.3 Types

0.3.1 Natural cork

Slabs of cork are steam treated to flatten them in order that they can be laminated to any size and thickness. As this is virtually pure cork there is no degeneration.

The reason that cork is so perfect for this application is its ability to absorb sound and vibration, to compress and then to recover, whilst essentially remaining unaltered. It will readily accept coatings which are sometimes used to protect adjacent materials from extraneous matter.

0.3.2 Agglomerated corkboard

Agglomerated corkboard is in sheet form, usually 1000 mm × 500 mm × 50 mm, or as required. Corkboard for this purpose should be of a minimum density of 176 kg m^{-3}. It is manufactured by the same method as insulation corkboard and looks similar. Unlike the other types, this material, being of lighter density and lower tensile strength, is usually laid within a concrete base.

0.3.3 Composition cork

Sheets of composition cork are usually 914 mm × 610 mm × 5 mm/50 mm. This type of material is made up of smaller granules than corkboard. A binding agent is added and the material is then subjected to heat and pressure. The grades which are suitable for this purpose should be between 440 kg m^{-3} and 480 kg m^{-3}.

0.3.4 Cork/rubber

Cork/rubber sheets are usually 914 mm × 914 mm, 1000 mm × 500 × 12 mm/25 mm, or as required. This is a mixture of cork granules and synthetic rubber binder which has been subjected to heat and pressure. There are a number of grades to suit the particular requirement of the base. Typical properties are shown in *Table 10.1*. Physical characteristics of insulation corkboard are shown in *Table 10.2*. Standard thicknesses of corkboard are shown in *Table 10.3*.

0.4 Applications

Expanded corkboard is divided into four groups: thermal, acoustic, anti-vibration, and decorative. Experiments have shown that cork is unique in having a high performance in all these categories.

The optimum material combines a low density with the greatest mechanical and structural strength. This combines advantageous technical characteristics with a competitive price, although a material with a higher density can be perfectly acceptable. For this reason insulation corkboard cannot effectively be specified in terms of density alone.

The ideal quality is achieved by careful selection of raw materials, eliminating the heavier woody inner bark sometime found in the denser grades of corkboard.

In this context it is interesting to note that, at the ISO TC/87 (cork) Plenary Session at Merida in 1987, the importance of the physical properties of corkboard relative to its density was recognized and it was decided to issue a Draft Revision of the existing specification for comment and further discussion taking this factor into account.

However, it is recommended that any material that is less than $96\,kN\,m^{-2}$ and greater than $140\,kN\,m^{-2}$ be given extra scrutiny to ensure that it complies with the other vital criteria

Cork is particularly user-friendly for the following reasons:

(1) Compressibility
(2) Softness
(3) Non-abrasiveness
(4) Grip on surrounding materials
(5) Ease of removal
(6) Ease of drilling
(7) Environmentally friendly.

10.5 Standards

International Standard ISO 2219 embraces all the major requirements as follows:

Dimensional tolerances:
on length $\pm 3\,mm$
on width $\pm 1.5\,mm$
on thickness
up to 25 mm thick $\pm 5\%$
25–50 mm thick $\pm 3\%$
> 50 mm thick $\pm 2\%$.
Density: $< 140\,kg\,m^{-3}$
Modulus of rupture: $> 140\,kN\,m^{-2}$
Thermal conductivity: $< 0.042\,W\,m^{-2}\,°C^{-1}$ (at 20°C); however at 0°C, max. $0.037\,W\,m^{-2}\,°C^{-1}$.
Moisture content: max. 3% by volume.
Compressive strength: at 10% compression, $120–140\,kN\,m^{-2}$

The methods of determination of these standards are also covered by ISO specifications.

The critical properties are that density should be kept as low as is practical by using the best quality corkwood to give acceptable modulus of rupture and compressive strength figures. Poor quality material is friable and prone to disintegration and may consist of large granules with many voids or heavy woody bark.

Table 10.1 Variation of properties with specification

Specification reference	Cos46	Cos40	Cos70	712BNF	COR12V
Type	Synthetic	Neoprene	Synthetic	Nitrile	Neoprene
Hardness, shore A	70–80	65 ± 10	65 ± 5	70–85	65 ± 5
Tensile strength	200	250 min.	200 min.	400 min.	300
Grade	Medium	Medium	Medium	Firm	Firm
Flex	5 max.	Pass 3	3 max.	3×	3 max.
Specific gravity	0.71–0.75	0.63–0.72	0.56–0.72	0.80–0.88	0.85–0.95
Compression at 400 psi	25–35%	25–35%	35–45%	15–25%	–
Recovery	75% min.	80% min.	80% min.	80% min.	97.3% min.
Volume change (%)					
Oil 1	–5 to +15	0 to +5	–5 to +5	–2 to +20	–
Oil 3	0 to +20	+10 to +30	–2 to +15	+15 to +50	–
Fuel A	0 to +15	0 to +10	–2 to +10	0 to +15	–
Water absorption	–	–	–	–	–
Compression set A (ASTM)	–	–	–	–	–
Compression set B (ASTM)	–	–	–	–	–

Table 10.2　*Physical characteristics of insulation corkboard*

Property	Value	Standard
Density	104–128 kg m^{-3}	DIN 18–161, ISO 2189
	109 kg m^{-3} average	British Board of Agrement*
Thermal conductivity	At 0°C, 0.035–0.037 W m^{-1} K^{-1}	DIN 52–612, ISO 2502
	At 20°C, 0.42 W m^{-1} K^{-1}	British Board of Agrement*
Compressive strength	At 10%, 0.25 N mm^{-2}	DIN 18–161
Deformation under temperature	2.4% using 30 mm corkboard, 80°C	DIN 18–161
	1.4% using 50 mm corkboard, 80°C	DIN 18–161
Coefficient of thermal expansion	35 × 10^{-3} mm m^{-1} measured between 20° and 100°C	
Specific heat	1.67–2.09 kg K^{-1}	
Thermal diffusivity	0.00067 m^2 h^{-1}	
Spread of flame	Subject to coverings	
	Ext. F.A.A.	Yarsley Technical Centre (Indicative)
	Ext. F.A.B.	Warrington Test Laboratory (Indicative)
Behaviour in fire	Does not liberate toxic gases	
Working temperatures	−180°C to 110°C	
Building regulations	Deemed to satisfy material	Regulation 7. Clause 1:2
Compatibility	Suitable for fully bonded bitumen specifications	
	Suitable for use under mastic asphalt	
	Suitable for single-ply fully adhered membranes	Consult Technical Literature
Wind resistance	2.5 kN m^{-2} fully bonded +2.5 kN m^{-2}, additional loading layer or mechanical fastenings required	

Vermin and rot proof, does not encourage growth of fungi, mould or bacteria.
A remarkably strong, stable, resilient and safe insulant for flat-roof installations.
Insulation corkboard has been proven in service for over 50 years and will give
effective protection for the life-time of the waterproofing layers.

Recovery: loading test of 4 s duration with 0–140 kN m^{-2} load.
Completed 1200 times, the maximum loss of thickness is
0.12 mm of the 60 mm sample tested.

The thermal diffusivity (a measurement of heat flow calibrated
against time) shows that cork is both a good insulant and
resistor to changes in temperature and can be used in conjunction with oil-based products to improve their performance.
Further, the recovery of cork after loading is significantly
greater than polystyrene and many other oil based products.

10.5.1 Dimensional stability

The dimensional stability of cork is 3.4×10 (BOC), which
means an expansion of 0.03 mm m^{-1} for a temperature rise of
1°C which would give a maximum expansion range in our

Table 10.3 *Standard thicknesses of insulation corkboard*

	Thickness (mm)							
	150	100	80	60	50	40	30	20
No. of square metres per pack/carton. Available standard density (104–125 kg m⁻³)	1.0	1.5	2.0	2.5	3.0	4.0	5.0	7.5
Weight (kg m⁻²) (density 109 kg m⁻³)	16.35	10.90	8.72	6.54	5.45	4.36	3.27	2.18
U at 0°C	0.23	0.34	0.42	0.53	0.63	0.75	0.95	1.20
U at 20°C	0.26	0.38	0.47	0.60	0.69	0.84	1.05	1.34
Metal deck-maximum trough dimension. Available in heavy density (170 kg m⁻³)	150	125	125	100	100	75	75	50

Full product range includes: other sizes available to special order; prefinished boards for PVC membranes available to order; high density fillets; specific cutting to individual project requirements; and tapered roof specialists—design and production.

climate of $0.9\,\mathrm{mm\,m^{-1}}$. This can be compared with the value for polyurethane which is many times greater.

High performance insulation must not only provide containment of warmth (or cold) but must not react violently to changes in internal and external conditions, especially when differing thicknesses are used on tapered insulation schemes.

Corkboard can, therefore, be described as 'benign', making an ideal companion to other components.

10.5.2 Fire

In considering the performance of a material in a fire it is necessary to look at the roof as a whole including the deck and membrane. Therefore there is little that can be said about a material in isolation but, most importantly, cork does not emit toxic fumes and satisfies Class 2 of BS 476 (1971) as a surface with a very slow propagation of flame. Cork can therefore be considered to have good fire resistance.

10.5.3 Quality

It is particularly relevant that suppliers work to quality systems that are to the standard of BS 5750 *Quality systems*, 1982–87.

10.6 Site precautions

10.6.1 Handling and storage

Insulation corkboard is light in weight and may be cut using conventional tools. It should be handled with care to avoid damage to board edges. The slabs are shrink wrapped in polythene or corrugated export cartons which will provide short term protection. For long-term protection they should be stored indoors or off the ground under a secure waterproof covering.

10.6.2 Fixing

Corkboard is normally bonded with hot bonding-grade bitumen, using a staggered joint arrangement. A vapour check barrier should always be installed below the insulation and the specification carefully chosen to suit the individual project requirements. Day work joints should be fully sealed to prevent ingress of moisture. Additional mechanical fastenings may be required for roofs in highly exposed locations. Care should be taken to follow the manufacturer's technical instructions.

10.6.3 Roof finishes

10.6.3.1 Built-up proofing A vapour escape sheet/partial bonding layer is not recommended or required with cork board. High performance built-up roof should be fully bonded to corkboard. The use of elastomeric membranes is recommended, although not mandatory.

10.6.3.2 Single-layer high-performance membranes Corkboard is suitable for use with polyvinyl chloride (PVC) membranes, but it is recommended that the product is supplied with a sealed finish compatible with the adhesives to be used. The manufacturer's technical instructions should be followed carefully.

10.6.3.3 Asphalt No thermal barrier layer is required. A loose laid standard isolating membrane should be laid over the corkboard prior to application of asphalt to MACEF (Mastic Asphalt Council Employers' Federation) recommendations.

Bibliography

OLLEY, R. In DORAN, D.K. (ed.), *Construction Materials Reference Book*, Butterworth-Heinemann, Oxford, Chapter 26 (1992)

11 *Fabrics*

Woven fabrics have a long history. First dominated by natural fibres from plants, cellulose in the form of flax and cotton, and from animals, e.g. silk and wool, the initial uses of fabrics were in the manufacture of tents, sails and ropes for both domestic and military boat and shelter uses. However, these structures were restricted to a relatively small scale since their inherent fibres were highly subject to creep, rot and ultraviolet (UV) degradation.

It is only during the 20th century that fabrics have been specially designed to fit machines and structures able to bear variable loadings arising mainly from snow, wind and temperature. The development of airships provided the earliest starting point of research on fabric properties and the determination of stress–strain behaviour. More recently, in the late 1950s, the conjunction of artificial composites with the application of the surface-curvature principle stiffened by prestressing, made it possible to use fabric to meet the architectural requirements of reliably covering areas larger than those of conventional towing tents or circuses.

A predictable performance has been brought to fabric structures due to the fundamental change that has arisen from recent developments in coated fabrics, which resist degradation, and from improvements in structural forms and detailing which allow the full potential of these materials to be utilized and justified. The design of stressed-skin structures may utilize individual ropes or cables knotted or clamped together (net), and any of a range of fabrics, from coated fabrics to film or sheet materials. By far the largest-scale applications use coated fabrics for which a wide range of fibres and coating materials have now been studied.

A list of fibres used as the basic components of structural cloths is given in *Table 11.1* and a list of coating materials is given in *Table 11.2*. It is important to note that not all coatings can be applied to all base fibres. For instance, polytetrafluoroethylene (PTFE) and FEP have fairly high melting points and can, therefore only be coated to high-temperature-resisting fibres. The main point to note, however, is that amongst the numerous compatible combinations of cloth and coating, only three are readily commercially available. These three composites are:

(1) Polyvinylchloride coated polyester (PVC/PES)
(2) PTFE-coated glass (PTFE/GS)
(3) Silicone-coated glass (VESTAR).

Table 11.1 *Fibres used as the basic components of structural cloths*

Natural fibres
Cotton
Flax

Synthetic fibres
Polyethylene
Polyamide
Polypropylene
Viscose
Extended-chain polyethylene
Polyester tergal
Aramid, high and low modulus

Metal fibres
Steel
Copper alloys

Mineral fibres
Glass E
Glass S
Carbon
Graphite
Bore

Table 11.2 *Coating materials used on structural cloths*

PVC
Polyurethane
Polychloroprene
Natural rubber
Hypalon (chlorosulphonated polyethylene)
PTFE and FEP (Teflon)
Acrylic
PVF
PVDF
Silicone rubber

PVF, polyvinyl fluoride; PVDF, polyvinyl difluoride.

Table 11.3 *The breaking strength versus elongation of some fibres*

Fibre		Breaking strength (hbar)	Elongation (%)
1	Carbon	210	0.3
2	Steel	350	2.0
3	Aramid	290	2.0
4	Glass S	350	3.2
5	Glass E	220	2.5
6	FCPE	270	3.8
7	FAX	90	2.0
8	Rayon	145	10.0
9	Cotton	70	6.0
10	Polyester	150	16.0
11	Nylon	115	21.0
12	Acrylic	50	24.0

Table 11.4 Cost, properties and applications of fabrics

Material	Cost per m² (£)	ΔL/L* (%)	Durability (years)	Translucency (%)	Fire resistance†	Colour range	Application
Foils							
PVC	1.5		<10	90	None	All	
Polyester MYLAR	5.0		<15	95	None	All	
FEP	10	200	<25	95	M1	Clear	Green-house glazing
PETFE	10	300	>25	90	M1	Clear	
Coated fabrics							
Polyester/PVC with acrylic lacquer	2–7	16	<15	8–30	M2	All	Widely used for all types of tensile structures
Nylon/PVC	2–7	20	<12	8–30	M2	All	Poor dimensional stability
Kevlar/PVC	30–60	3	<20	Opaque	M2	All	Only used where high strength is required
Polyester/Hypalon	8–20		<20	Opaque	M0	All	Used for radar domes
Polyester/PVC with Tedlar	3–8	16	<20	7–20	M2	All	Better self-cleaning ability; little experience on welded joints
Polyester/PVDF	15–20	16	<25	35	M2	White	In development

Table 11.4 Continued

Material	Cost per m² (£)	$\Delta L/L^*$ (%)	Durability (years)	Translucency (%)	Fire resistance†	Colour range	Application
Glass/PTFE	25–45	6	<30	5–15	M0	Ivory	Widely used in USA for permanent membrane
Kevlar/PTFE	40–65	3	<25	Opaque	M1	Ivory	Very seldom used
Glass/silicon	20	6	<25	20–50	M2	Clear	New material used for 5 years in USA
Reinforced films							
PVC/polyester	1		<10	80	M2		Widely used for clear sheeting on scaffolding
FEP or ETFE with glass, Kevlar or steel-wire mesh			<25	50–80			Not in commercial production

*Relative strain.
†M0, M1, M2. French Regulations.

Fibres are available in a range of breaking strengths. *Table 11.3* and *Figure 11.1* give data on the breaking strength versus elongation of several different fibres. *Table 11.4* indicates cost, further properties and applications for structural fabrics.

Several experimental studies and observation of the real-time behaviour over the past 20 years of polyester-PVC fabrics have shown that the main factor affecting the durability of these materials over long time exposure is the thickness and plasticity of the PVC coating used for ultraviolet-radiation insulation. However, for PTFE-coated glass, the waterproofing of the coating is the most important factor since glass fibres are weakened by water exposure.

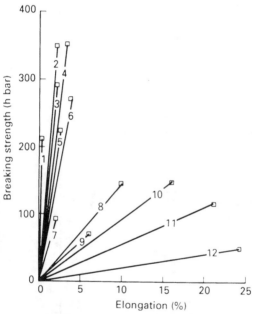

Figure 11.1 *Plots of breaking strength versus elongation for the fibres listed in* Table 11.3.

Bibliography

MALINOWSKY, M. In DORAN, D.K., *Construction Materials Reference Book*, Butterworth-Heinemann, Oxford, Chapter 27 (1992)

12 *Glass*

12.1 Introduction

Very little glass is found in nature, although some does occur near volcanoes where sufficient heat has been generated to cause materials to fuse together. Obsidian—natural glass—is usually impure and, therefore, lacks transparency. Primitive peoples used it to make into arrow heads, knife blades, ornaments and other tools. Such objects have been found worldwide, for example in Africa, Greece, Australia, and Mexico.

The earliest examples of man-made glass have been found in the remains of early Middle Eastern civilizations in the form of beads and small vessels, dating back some 4000 years. Glass-making techniques spread gradually from Egypt and by the 6th century BC had reached most parts of the western Mediterranean. During the last three centuries BC, glass-makers in Alexandria perfected the technique of making composite coloured glass canes and rods. These were made by drawing together a number of glasses of different colours. The rods were then cut into slices to provide repetitive designs. By the time of Christ, the Romans had spread the art of glass-making throughout the Empire and they appeared to have mastered all the technical processes involved in glass-making and decorating. The collapse of the Roman Empire caused a decline in glass-making in Europe until the Venetians revived it around 1200 AD. During the intervening period the techniques for casting stained glass and the making of stained-glass windows were developed in Lorraine and Normandy.

Whilst the Venetians rediscovered all the skills exhibited by the Romans, it is believed that these skills spread not from Rome but from the earlier industry in Syria. By the beginning of the 14th century, Venetian craftsmen had mastered, for example, the production of fine glass mirrors. By 1600 the Venetian monopoly had eroded and the techniques had spread across Europe to Bohemia, France, England and the Netherlands.

Whilst the Venetians were concentrating on mirrors and glasses for decorative purposes, the glass-makers of Lorraine and Normandy were developing methods of manufacturing flat glass. The cylinder drawn and crown processes were introduced into Britain in the 15th and 16th centuries.

The location of glass-making enterprises was usually determined by the presence of wood for fuel, and streams from which sand was obtained. Alkali was normally obtained by burning bracken or seaweed. In England most of this early activity was centred in the south of England, the Weald in

Kent and Sussex being the most famous. Towards the end of the 16th century wood supplies became seriously depleted and in 1615 the use of wood for glass-making was banned by law. The ban on the use of wood encouraged the changeover to coal and with it some important changes in furnace design.

The adoption of coal was a further major influence on the location of the glass-making industry, especially as at this time the canal and railway systems had not been developed and, therefore, as well as the location of raw materials, transportation was a major factor. In the UK the four main areas of activity were Newcastle, Bristol, Birmingham and St Helens. Over the years there have been many developments in the basic manufacture of flat glass for architectural applications until at the present time the major glass-making process used is the float glass process developed by Pilkington in 1959. In the Western World over 90% of flat-glass production is by the float process.

Glass in Building (Ed. D. Burton and B. Pye, Butterworth Architecture, 1993) includes a great deal of information on glass.

The manufacture of all glass products is based on four fundamental stages:

(1) Melting
(2) Forming
(3) Cooling
(4) Finishing.

A typical composition for architectural glass is as follows:

> sand (silica) 72%
> soda ash (sodium carbonate) 13%
> limestone (calcium carbonate) 10%
> dolomite (calcium magnesium carbonate) 4%
>
> (note % by weight).

These raw materials, weighed and mixed in the correct proportions, produce a mixture known as frit. Waste broken glass (cullet) is re-cycled along with frit, usually in the proportions of 80% frit and 20% cullet.

12.2 Types of glass

These can generally be classified as follows:

(1) Float glass
(2) Sheet glass
(3) Rolled glass
(4) Wired glass
(5) Body coloured/tinted glass
(6) Surface modified glass

(7) Processed glasses—including coated; toughened; laminated; silvered; other surface treatments; insulating; blocks and fire resistant.

12.2.1 Float glass

As the name implies this is made using the float process. The width of sheet is usually up to 3 m, the length being limited only by handling considerations. The approximate composition (by weight) is:

Silicon	SiO_2	70–74%
Lime	CaO	5–12%
Soda	Na_2O	12–16%
Magnesium	MgO	0–5%
Aluminium	Al_2O_2	0.2–2%

Other trace elements may be present. The following properties are typical of clear float glass. Minor variations between manufacturers may occur:

Density	$2560\,kg\,m^{-3}$
Hardness	6.5 Mohs
Young's Modulus	$74.5 \times 10^9\,Pa$
Poisson's Ratio	0.23
Specific heat	$830\,J\,kg^{-1}\,K^{-1}$
Thermal conductivity	$1.05\,W\,m^{-1}\,K^{-1}$
Thermal transmittance	$5.68\,W\,m^{-2}\,K^{-1}$
Sound transmission	

Thickness (mm)	Sound insulation (dB) (mean 100–3150 kHz)
4	25
6	27
8	29
10	30
12	31

Coefficient of linear expansion	$7.6 - 8.0 \times 10^{-6}\,K^{-1}$
Refractive index	1.52 (380–760 mm)

The light transmission and total-solar-energy transmission of clear float glass is shown in *Table 12.1* for a range of common thicknesses.

12.2.2 Sheet glass

This type of glass is dying out and being replaced by float glass as new plants come on stream. Most true sheet glass is manufactured in Eastern Europe and Third World countries. Because of the method of manufacture the surface quality is

Table 12.1　*Light and total transmission of clear float glass*

Thickness (mm)	Light transmission (%)	Total solar energy transmission (%)	Normal maximum size* (mm)
3	88	83	2140 × 1220
4	87	80	2760 × 1220
5	86	77	3180 × 2100
6	85	75	4600 × 3180
8	83	70	6000 × 3300
10	84	65	6000 × 3300
12	82	61	6000 × 3300
15	76	55	3050 × 3000
19	72	48	3000 × 2900

*These sizes vary with manufacturer and for precise data individual manufacturer's literature should be consulted.

more variable than that of float glass. Sheet glass is usually graded by visual inspection into the following:

OQ　ordinary quality
SQ　selected glazing quality
SSQ　special selected quality.

The light and total energy transmission factors are given in *Table 12.2* for a range of common thicknesses.

12.2.3 Rolled glass
This designation covers all glasses which are passed between rollers to give a specific surface finish, whether it is a definite regular pattern or a random diffusing surface. Rolled glasses are usually used where privacy or diffusion is required or where

Table 12.2　*Light and total solar energy transmission of clear sheet glass*

Thickness (mm)	Light transmission (%)	Total solar energy transmission (%)	Normal maximum size* (mm)
3	88	83	2140 × 1220
4	87	80	2760 × 1220
5	86	77	3180 × 2100
6	85	75	4600 × 3180
8	83	70	6000 × 3300
10	84	65	6000 × 3300
12	82	61	6000 × 3300
15	76	55	3050 × 3000
19	72	48	3000 × 2900

*See footnote to *Table 12.1*.

specific decorative effects are required. There is a wide range of rolled patterns to choose from, each manufacturer offering a pattern range. For details it is really necessary to consult the manufacturers' literature in order to establish what is suitable for a particular application.

The depth to which the pattern is rolled into the glass and the average thickness determines the extent to which rolled patterned glasses can be laminated or toughened. Again manufacturers' literature must be consulted to establish the suitability of a particular pattern for these processes.

12.2.4 Wired glass

Whilst not usually classified as a true safety glass, in the event of bodily impact (in the various codes of practice dealing with safety), the presence of the wires does hold the glass fragments together when breakage occurs. It is this characteristic that has resulted in wired glass being used in large quantities in overhead glazing applications. However, the main use of wired glass has always been to create fire-resistant-glazing systems. In such installations the wire maintains the integrity of the glazed panel which, if glazed properly, can achieve fire-resistance ratings of 60–90 min.

Manufacturers' literature should be consulted for precise details of fire ratings and sizes which are acceptable. The usual thicknesses available are 7 mm for rough cast or patterned wired glass and 6 mm when in polished form. The maximum sizes are in the range 3700 mm × 3300 mm to 1840 mm × 1830 mm. On the whole, polished examples tend to be the smaller sheet sizes.

12.2.5 Body coloured/tinted glass

Body coloured and tinted glasses are produced by varying the constituents in the melting furnace. The colour is present throughout the thickness of the glass with the result that different thicknesses of the same mix will have different light and total solar energy transmissions. Thicker examples will appear more strongly coloured, which has important implications for the mixing of glass thicknesses on the same facade of a building. Body coloured and tinted glasses are used either as solar-control glasses where changes to the glass composition increase the absorption of the material and reduce the amount of light and total solar energy transmitted. So-called tinted glasses are used partly for decorative purposes.

Because the manufacture of these glasses can only be achieved by the lengthy process of changing the basic composition of the glass mix they are being superseded by other types of manufacturing processes which rely on modifying the performance and appearance of basic clear float glass. Body coloured and tinted glasses are normally produced in green, grey and bronze forms, although other tints are available. The individual manufacturer's literature should be consulted for details

Table 12.3   *Performance data for some typical body tinted glasses*

Type	Thickness (mm)	Light transmission (%)	Total solar energy transmission (%)	Normal maximum size* (mm)
Body tinted green float	6	72	62	3210 × 6000
Body tinted bronze float	4	61	70	3210 × 6000
	6	50	62	3210 × 6000
	10	33	51	3210 × 6000
	12	27	47	3210 × 6000
Body tinted bronze float	4	55	68	3210 × 6000
	6	42	60	3210 × 6000
	10	25	49	3210 × 6000
	12	19	45	3210 × 6000

of colours, sizes and performance data. *Table 12.3* gives some typical data.

12.2.6 Surface-modified glass

Surface-modified glasses are usually produced on the float line as a continuous process and are achieved, as the name suggests, by modifying the surface of basic clear float glass. In some instances the modification occurs within the float bath whilst the glass is still in a semi-molten state. In other cases the modification occurs whilst the glass is still hot, but is at a convenient stage in the annealing cycle. Surface modification within the float bath is achieved either by bombarding the semi-molten glass with metal ions or by chemical vapour deposition. When the modification is performed during the annealing cycle the modifying process is usually referred to as a pyrolitic coating since it involves the pyrolitic decomposition of materials sprayed onto the glass surface. The modifying layer fuses with the glass surface.

Surface modified glasses are normally used for solar control or for applications where low-emissivity characteristics are required. A wide range of performance characterstics is available as a result of surface modification. The individual manufacturers' literature should be consulted to establish the range of performance, appearance and sizes available.

12.2.7 Processed glasses

12.2.7.1 Coated glass   Various processes have been developed for the application of thin coatings to large areas of flat glass. Coatings are applied to alter the performance of the basic float glass, i.e. light and radiant-heat transmission, light and radiant-heat reflection, and surface emissivity. Coatings which have been developed to have high transparency but high reflectivity of long-wave radiation (low emissivity coatings) are used to improve the thermal-insulation properties of the treated glass.

More than one coating can be applied to the glass surface to give particular properties. Coatings can be applied to glasses for incorporation in sealed-double-glazing units and for laminating. Some coatings provide distinctive colours to the glass, whilst others are neutral in appearance. Many major glass manufacturers have coating facilities, as do glass processors.

The coatings used can be divided in two groups: metallic coatings and metallic oxide coatings. Metallic or metallic alloy coatings are applied to the glass by means of a vacuum deposition process, whilst coatings involving metallic oxide can be applied in a normal atmosphere. The application of an oxide coating requires temperatures of around 500°C which in effect 'fire' the coating to the base glass.

Metallic oxide coatings tend to be hard coatings which are sufficiently robust to withstand exposure to the normal atmosphere and weather. Metallic and metallic alloy coatings are usually softer and may not be sufficiently resistant to scratching and weathering without protection. Metallic and metallic alloy coatings are normally incorporated into double-glazing units or laminated products.

The use of gold, copper and silver coatings not only provides good solar protection and range of colour effects but also improved thermal insulation properties. Significant improvements in U values can be achieved by a combination of coated glass and cavity construction. A unit comprising two pieces of uncoated float glass and a 12 mm air space would typically have a U value of $3.0 \, \mathrm{W \, m^{-2} \, K^{-1}}$. The use of two coated glasses could reduce the U value to $1.6 \, \mathrm{W \, m^{-2} \, K^{-1}}$ with further improvement down to 1.4 if the air in the cavity is replaced by an inert gas such as argon or krypton. Further data on coated glass is shown in *Table 12.4* and typical partition-of-energy diagrams are shown in *Figure 12.1*.

12.2.7.2 Toughened glass Toughened glass is usually produced by reheating a piece of annealed glass to a temperature of approximately 700°C at which point it begins to soften. The surfaces of the heated glass are then cooled rapidly.

The technique creates a state of high compression in the outer surfaces of the glass and, as a result, although most other

Table 12.4 *Typical properties of reflective coated glasses*

Coating	Light transmittance (%)	Total solar transmittance (%)	Solar reflectance (%)
Gold	42	22	40
Copper	42	25	39
Silver	42	27	47

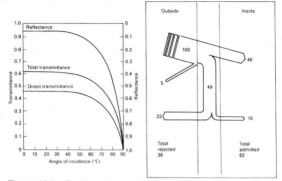

Figure 12.1 *Partition of energy by a 6 mm 'Antisun' float 50/62 (bronze)*

characteristics remain unchanged, the tensile strength is usually increased by a factor of four or five times that of annealed glass. When broken the toughened glass fractures into relatively small pieces usually in the form of small cubes. The lengths of the sides of the cubes are normally the same as the thickness of the glass. Because cubes of broken toughened glass do not have the sharp blade-like edges and dagger points of broken annealed glass it is regarded as a safety glazing material in the various codes of practice and standards.

Toughened glass is usually produced by one of two main methods based on heating followed by rapid cooling. In one method the glass is held vertically by tongs along the upper edge and suspended in a heating furnace to raise its temperature to about 700°C. The glass is then cooled rapidly by placing it between an array of nozzles which blast cold air onto the surfaces and cool the glass rapidly. The other process used involves supporting the glass horizontally on rollers and first passing it into the heating chamber then into the cooling area. This is known as the roller hearth process. Both processes introduce some slight distortion of the glass, and the former also causes pinch marks where the tongs have gripped the glass.

All float and sheet glasses can be toughened. Many rolled patterned glasses can also be toughened depending on the profile of the patterns. Wired glasses cannot be satisfactorily toughened. The glasses which are toughened must be cut to size and have any other processing such as edge polishing and hole drilling completed before being subjected to the toughening process. Any attempt to 'work' the glass after toughening will cause the glass to shatter.

Whilst the air toughening process is being used it is possible to coat the glass with a coloured ceramic material which is then fired into the surface of the glass during the heating cycle of the toughening process. This results in a coloured opaque cladding glass which can be used in spandrel areas of buildings.

Whilst toughened glass is stronger than annealed glass of the same thickness it is more normal to use it in applications where a safe breaking pattern is required rather than in situations where additional strength is required.

The toughening process used places limits on the maximum sizes which can be handled. Typical maximum sizes are listed in *Table 12.5*.

12.2.7.3 Laminated glass Laminated glass is produced by bonding two glasses together with a plastic material or a resin. The interlayer, which is usually polyvinyl butyral, can be either clear or tinted. The bonding is achieved by heating the glass/interlayer sandwich and applying pressure.

When a resin is used as the bonding medium a self-curing resin is usually poured between the two pieces of glass which are maintained at the correct separation whilst pouring and curing takes place. This process which is often referred to as 'cast in place' is more appropriate to small-scale production of special glasses.

Laminates can incorporate several thicknesses and combinations of glasses. Different thicknesses of interlayer can be specified to give a selection of products with a wide range of properties and applications. Laminated glasses can be used as safety glazing, bullet-resistant glazing, glazing for sound attenuation, solar control, vandal resistance, and for overhead applications.

When a laminated glass is broken the interlayer tends to hold the fragments of broken glass in place. The actual breakage pattern is the same as for annealed glass, although the strength of laminated glass is slightly less than that of monolithic annealed glass of the same substance.

Laminated glasses can incorporate multiple layers of interlayer and glass plus other material such as polycarbonate to achieve specific performance characteristics. Toughened glasses can also be used as components of a laminated product.

12.2.7.4 Silvered glass Silvering is a chemical process normally used to create mirrors by depositing a layer of metallic silver onto the surface of a piece of clear glass. The silver deposit is usally protected by a layer of copper which in turn

Table 12.5 *Typical maximum sizes for toughened glass*

Thickness (mm)	Size (m^2)
4	2300 × 1300
5	2600 × 2000
6	4200 × 2000
10	4200 × 2000
12	4200 × 2000
15	4000 × 1800
19	4000 × 1800

is protected by a special paint coating. The silver surface which gives the mirror its reflective properties is viewed through the glass. As an alternative to plain silver, special effects can be created by using tinted glass or by colouring the deposit of metal.

For special purposes 'front silvered' mirrors can be produced. These have the reflective material on the front surface of the glass. They are not normally used in architectural applications since the silvered surface has limited durability without additional protection.

It is also possible to create mirrors by using metals like aluminium to create the reflective surfaces.

12.2.7.5 Other surface treatments   These include

(1) Sand blasting
(2) Acid etching
(3) Screen printing
(4) Engraving
(5) Brilliant cutting
(6) Enamelling
(7) Staining
(8) Painting.

For further information contact specialist processors.

12.2.7.6 Bent glass   This can be produced by heating most basic glasses to the point where they soften until they can either be pressed or sag bent over suitable formers. Bends can be created in two planes.

12.2.7.7 Insulating glass   This incorporates two or more glasses separated by spacers to create a cavity between successive panes to form a unit. Spacers are attached by suitable adhesives; cavities normally vary between 5 and 20 mm. Typical thermal transmittance values are shown in *Table 12.6*.

12.2.7.8 Glass blocks   These are made in a variety of patterns, sizes, colours and degrees of transparency. They are

Table 12.6   *Typical thermal transmittance (U values)**

Airspace width (mm)	U value (W m^{-2} K^{-1})
6	3.4
8	3.2
10	3.1
12	3.0
16	2.9
20	2.8

*Air in cavity; two panes of 4 mm glass.

extensively used in the USA and continental Europe; less so in the UK. These units are usually laid in a manner similar to that used for normal building bricks using a fine sand : lime : cement mortar. *Table 12.7* provided by Luxcrete Ltd, gives typical details of available patterns, sizes and quantities per square metre.

12.2.7.9 Fire-resistant glasses There are several types of fire-resistant glass. These can provide up to 60 min of stability/integrity when correctly framed. The following approaches are relevant:

(1) Wired glass
(2) Laminated float glass with gel interlayers
(3) Prestressed borosilicate glass
(4) Toughened calcium/silica float glass
(5) Glass blocks.

Technical data is available from specialist manufacturers.

12.3 Structural glazing

To some extent most glazing is used in a structural manner and it is important to use the appropriate specification, thickness and bedding technique to resist the forces to be resisted. Guidance on some of the loads to be resisted can be obtained from BS 6399: *Loading for buildings Part 2: Code of practice for wind loading*.

Systems have been devised which allow toughened glass plates to be assembled to create continuous glass facades without the use of mullions and transoms. The plates are usually bolted together using special metal patch fittings at

Table 12.7 *Data for glass blocks. (Courtesy Luxcrete Ltd, London NW10 7BT)*

Pattern	Reference	Size (mm)	Blocks per square metre
Flemish	L.190F	190 × 190 × 80	25
Cross reeded	L.190CR	190 × 190 × 80	25
Clear	L.190C	190 × 190 × 80	25
Flemish	L.240F	240 × 240 × 80	16
Cross reeded	L.240CR	240 × 240 × 80	16
Clear	L.240C	240 × 240 × 80	16
Flemish	L.2415F	240 × 115 × 80	32
Clear	L.2415C	240 × 115 × 80	32
Flemish	L.115F	115 × 115 × 80	64
Clear	L.115C	115 × 115 × 80	64
Flemish	L.300F	300 × 300 × 100	9
Cross reeded	L.300CR	300 × 300 × 100	9
Clear	L.300C	300 × 300 × 100	9

their corners. In more sophisticated systems glass facades are given lateral stability by glass fins attached to the facade by suitable structural silicones. Where such techniques are envisaged it is strongly recommended that specialist advice is sought.

12.4   Cutting of glass

Cutting up to 10 mm thickness is usually achieved by scoring with a diamond or tungsten carbide point or wheel followed by snapping along the score line. For thicker glass it may be necessary to use a diamond-tipped saw. A variety of edge finishes (flat ground; round; bevelled or bull-nosed) may be achieved by grinding and polishing. Water-jet cutters may be used to achieve holes and irregular shapes.

12.5   Specification

The following are some of the items to be considered in compiling a specification:

(1) Legislation
(2) Codes and Standards
(3) Loadings (wind, fire, etc.)
(4) Security
(5) Architectural considerations
(6) Thermal and acoustic insulation.

Bibliography

JACKSON, G.K., In DORAN, D.K., *Construction Materials Reference Book*, Butterworth-Heinemann, Oxford, Chapter 29 (1992)

13 *Mineral-fibre products*

Mineral-fibre products have been made in the UK since the late 1880s, and in significant and increasing quantities for almost 50 years. From the early days, when relatively small quantities of crude mineral-fibre products were used almost exclusively as lagging for high temperature industrial plant, the industry has grown to a multi-billion tonne per annum, world-wide operation.

Man-made mineral-fibre products today include: continuous-filament glass fibres used primarily for reinforcement in a range of rubber, plastic and cement-based products; woven-glass cloths for the textile industry, small-diameter glass fibres for aircraft insulation and filtration purposes; ceramic fibres produced from refractory materials and used for very-high-temperature insulation applications; rock-fibre products used in horticulture as a growing medium and for general soil conditioning; and in their role as thermal and acoustic insulation.

It was the construction industry, however, that was primarily responsible for the significant increase in mineral-fibre production. The first major boost came in 1945 when, following the end of World War II, the demand for quickly erected, low cost, housing brought about a boom in lightweight prefabricated dwellings. Unlike traditional brick structures, this prefabricated construction provided no inherent thermal performance and, in order to improve living conditions, regulations were introduced requiring the inclusion of insulation in the outer skin of the dwelling.

At this stage most structural products still relied on the same bitumen bonding techniques which had been used since the turn of the century. In the mid-1950s, however, the formulation of special resin binders heralded a significant advance in manufacturing technology allowing increased production levels, more control over product characteristics, and vastly improved handling qualities.

Since the 1950s successive changes to the Building Regulations have steadily increased the levels of insulation required in buildings, initially for health and safety reasons and, with effect from the 1985 Regulations, to conserve energy.

'Mineral fibre' is a generic term encompassing all non-metallic inorganic fibres—a fibre being a particle with a length greater than 5μm and at least three times its diameter. Although mineral fibres occur naturally, for most modern-day applications the natural varieties have now been largely superseded by man-made alternatives.

Naturally occurring mineral fibres   Mineral fibres occur naturally as asbestos (see Chapter 6), a group of complex silicates which may contain iron, magnesium, calcium, sodium, etc., and which occur in veins as fibrous crystals. These crystals are unique in that they are capable of splitting longitudinally, when crushed, to form fibrils with diameters of much less than $1\mu m$ (typically 0.03–$0.04\mu m$).

Man-made mineral fibres   These are fibres manufactured primarily from glass, rock or other minerals, or from readily melted slags. In the construction industry, man-made mineral wool fibres are used primarily for thermal and acoustic insulation. The two most common types, which are generally referred to as 'mineral wools' because of their soft, woolly consistency, are:

(1) glass wool, a mineral wool (usually borosilicate glass) with a mean fibre diameter of 4–$9\mu m$
(2) rock wool, a mineral wool made from naturally occurring igneous rock (such as basalt or diabase) with a mean fibre diameter of 4–$9\mu m$.

Principal structural applications   Mineral-wool insulating products are available in rolls, batts and slabs for general use, preformed sections for pipe covering and loose wool for pouring or blowing applications. The products are used extensively for thermal and acoustic insulation in buildings, and for acoustic isolation of building elements (particularly floors). They provide thermal and acoustic insulation for heating, ventilating and air-conditioning systems and associated equipment, and are used in the manufacture of sound havens around noisy equipment and machinery. Mineral wools are particularly suitable for fire-stopping and general fire-protection applications, and special grades are available which meet the requirements for large and small cavity barriers.

13.1 Properties

13.1.1 Chemical composition
The percentage by weight of the main chemical components for typical glass and rock-based insulating wools are shown in *Table 13.1*.

The basic mineral wool is the predominant component of the insulating material, accounting for between 90 and 99.8% of the product by weight. Other components include:

(1) Phenol-formaldehyde resins (often modified with urea or lignosulphonates) 10% nominal
(2) Mineral oil, 1% nominal
(3) Silane, 0.02% nominal
(4) Polydimethylsiloxane (or water repellants based on this), 0.1% nominal.

Table 13.1 *Chemical composition of mineral-wool fibres*

Component	Content (wt%)	
	Glass wool	Rock wool
Silica (SiO_2)	53–65	45–54
Alumina (Al_2O_3)	2–5	8–15
Titania (TiO_2)	≤ 0.2	0.6–3
Ferric oxide (Fe_2O_3) ⎫	≤ 0.35	1.8–12.5
Ferrous oxide (FeO) ⎭		
Lime (CaO)	6.5–23.5	8–21
Magnesia (MgO)	2–5	3–14
Manganese oxide (MnO)	0.05	0.2
Sodium oxide (Na_2O)	9–18	1.6–16
Potassium oxide (K_2O)	0.8–1.7	0.4–1.6
Boric oxide (B_2O_3)	2–8	≤ 0.2

13.1.2 Factors affecting thermal conductivity

The relative levels of the various modes of heat transfer which work together to define the total thermal conductivity of the product, are affected by both the density and the temperature differential. The effect of density on thermal conductivity for 25 mm glass wool with a steady temperature differential of 20°C between the hot and cold faces is shown in *Figure 13.1*. This figure also shows the relative parts played by each of the heat-transfer modes under these very specific conditions.

The effect of temperature change on the thermal conductivity of a typical sample is shown in *Figure 13.2*. For normal structural applications, however, the effect of ambient-temperature changes on thermal conductivity are relatively small.

13.1.3 Sound absorption

Glass- and rock-wool products have excellent sound-absorbing properties and are particularly effective in reducing high-frequency noise. At thicknesses up to 50 mm the products are somewhat less effective at low frequencies, but improvements can be achieved by increasing material thickness where possible or, as is often the case, by including an air gap. The effect of increasing the flow resistance of mineral-wool products is to increase sound absorption in the middle- and high-frequency range. There is, however, an optimum relationship between air-flow resistance and sound absorption. Too high a flow resistance will lead to only a low friction heat loss, and both extremes will result in reduced sound absorption.

Although mineral-wool products provide good sound absorption when the surface is exposed, this is not always practical as often some form of protection will be required. In such cases it is possible to provide a porous membrane, or a very thin plastic film material (such as $12\mu m$ thick Melinex),

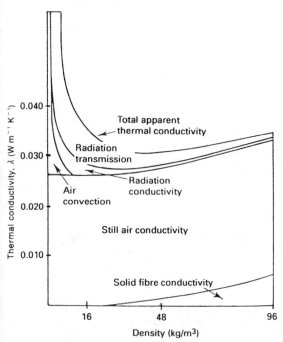

Figure 13.1   *Thermal conductivity vs. density*

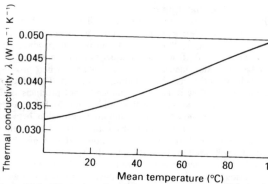

Figure 13.2   *Thermal conductivity vs. temperature for medium density mineral wool*

which only partially reduces acoustic absorption. Generally this will have the effect of reducing performance at high frequencies, although this may be offset by improvements in low-frequency-sound absorption.

13.1.4 Sound insulation

Sound insulation is concerned with the reduction of sound which passes between two spaces separated by a dividing structural element. Sound energy may be transmitted either directly through the element, or by indirect or flanking paths through the surrounding structure. In the case of walls and partitions, sound insulation is normally only concerned with airborne sound, but in floors impact sound transmission, such as that caused by footsteps, must also be considered in addition to airborne sound.

13.1.4.1 Airborne-sound insulation The resistance of any structural element to the passage of airborne sound is determined by measuring sound levels on either side of the element and recording the difference. This is known as the transmission loss. The level of sound insulation varies according to frequency and the results may be represented either graphically or in tabular form for frequencies at one-third octave intervals, generally over the range 100–3150 Hz. Standard test procedures defined in BS 2750 (ISO 140) are used to determine the sound insulation. An example of the performance of a typical lightweight partition is shown in *Figure 13.3*.

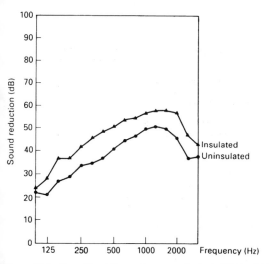

Figure 13.3 *Sound-insulation performance of a typical lightweight partition*

In single solid walls the sound-insulation performance is governed by mass—the heavier the material the greater its resistance to sound transmission. The empirical mass law makes it possible to predict the average sound insulation value of a solid structure of known mass. For example, a 100 mm thick solid block wall of average mass 100 kg m^{-2}, will have an average sound insulation of 40 dB (*Figure 13.4*).

Doubling the mass of the structure will generally provide an improvement of 5 dB (therefore doubling the wall from 100 to 200 mm thick would improve sound performance from 40 to 45 dB).

However, using lightweight-partition systems with mineral wool in the cavity it is possible to provide acoustic perfor-

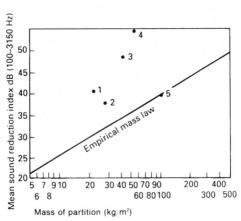

*Construction	Mass kg/m²	Mean R dB	Rw dB
1 75 mm metal stud partition with 25 mm mineral wool insulation in cavity	22	41	41
2 100 mm timber stud partition with 25 mm mineral wool insulation in cavity	27	38	40
3 100 mm metal stud partition with 25 mm mineral wool insulation in cavity	43	48	51
4 211 mm jumbo metal stud separating wall with 25 mm mineral wool insulation in cavity	55	55	59
5 100 mm solid block wall	100	40	40

Figure 13.4 *Sound insulation vs. mass*

mance better than that predicted by the mass law. For example, a typical 100-mm metal-stud partition with 12.5-mm plasterboard on both sides, but without insulation in the cavity, has a sound-reduction index of 35 dB. By adding 40 mm mineral wool in the cavity, the performance improves to 45 dB. With a total weight of 28 kg m^{-2}, the mass law would have predicted the performance of an equivalent solid partition as 32 dB. The improvement would have been even greater had the cavity been fully filled with mineral wool.

Another method of achieving good acoustic performance with solid walls is to use lightweight wall linings, such as mineral-wool/plasterboard laminates.

13.1.4.2 Impact-sound insulation The resistance of any floor/ceiling construction to impact-sound transmission is determined by a standard 'tapping' machine used in accordance with the procedures given in BS 5821: Part 2 (ISO 717: Part 2). In the test the noise levels in the lower room are measured in decibels at different frequencies, allowing a curve to be plotted. Using this method the lower the recorded level the better the acoustical performance of the system. *Figure 13.5* shows the performance of a typical concrete-floor construction when subjected to impact sound.

A mineral fibre quilt of suitable compressive resistance, inserted between a screed and the concrete floor, will improve impact sound insulation.

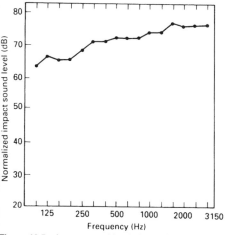

Figure 13.5 *Impact-sound insulation performance of a typical concrete floor*

13.1.5 Fire properties

Mineral-wool products perform well under fire conditions to frustrate both the passage of heat and the spread of flames. They are, therefore, widely used in buildings for both fire-protection and fire-control situations. The reasons for their superior performance are three-fold:

(1) The basic mineral wool is non-combustible, and does not add to the fire load of the building.
(2) The products have good thermal-insulation properties which slow and reduce heat transfer between the surfaces, giving protection to structural elements and preventing fire spread.
(3) The surfaces of the unfaced products act to resist the spread of flame.

The performance of mineral-wool products is unaffected at temperatures up to 230°C. At temperatures above 670°C softening of glass may commence. The breakdown process is one of softening rather than combustion; at no stage does any significant toxicity occur.

Bibliography

SMITH, P.M. In DORAN, D.K., *Construction Materials Reference Book*, Butterworth–Heinemann, Oxford, Chapter 30 (1992)

14 *Paints and preservatives*

14.1 Paints for non-absorbent surfaces

The non-absorbent materials of greatest interest to the construction industry are metals of which the most important is steel. This section, therefore, is concentrated on the protection and decoration of steel as a building material. Non-ferrous metals are frequently used for architectural fittings which may be coated with decorative paint and even stainless steel has occasionally required painting. Some of the high-performance paints applied to steel are also used on concrete floors and walls which are materials of low absorbency. Some reference will, therefore, be made to wall and floor treatment where there is an obvious connection with the coating types.

Iron and steel are reactive metals which when exposed to atmospheric oxygen and moisture rapidly convert to hydrated iron oxide in the form of rust. The rusting of iron and steel is an electrochemical process which takes place on the surface when adjacent areas form cathodes and anodes, and when a film of moisture bridges these areas to act as an electrolyte. Such conditions exist on structural steel when condensation takes place and the condensate contains ions, e.g. sulphate, sulphite, chloride, nitrate and nitrite from atmospheric pollution. These ions all serve to increase the electrical conductivity of the electrolyte.

Further information on the corrosion of steel may be found in Chapter 4.

One of the main causes of unsatisfactory coating performance is adhesion failure. The breakdown is obvious when a whole coating becomes detached from the substrate or when intercoat adhesion fails causing one coat to become detached from another in a multicoat system. The prevention of coating detachment can be avoided by careful attention to surface preparation.

14.1.1 Paint types
Paint consists of three distinct parts: the pigment, the binder and the solvent. Each of these parts may be a complex mixture of separate compounds.

Pigments The pigment is the insoluble part of the paint which gives it body. It may contain rust inhibitors, materials which absorb damaging ultraviolet radiation, hard particles to impart abrasion resistance and coloured material for appearance. When considering steel coatings it is convenient to class paints in terms of pigments being either 'inhibiting', i.e. containing rust-inhibiting properties, or 'non-inhibiting', i.e. paints used as a barrier or finishes having no rust-inhibiting properties.

Binders The binder, vehicle or medium in a paint is the film-forming part which is usually an organic resin. For paint used on structural steel it is convenient to consider binders under three classifications: single-pack convertible paints, single-pack non-convertible paints and two-pack chemical curing paints.

Single-pack convertible paints are based on drying oils or oil-modified resins.

Single-pack non-convertible paints dry by solvent evaporation only. Familiar examples are chlorinated rubber, vinyl co-polymers and acrylics.

Two-pack chemical curing paints are those to which a curing agent is added before application. Epoxies and polyurethanes are good examples of these materials which have excellent chemical resistance and durability.

Solvents Solvents are necessary in paint to make it flow and workable. Most paints for structural steel use organic solvents, although water-thinned primers are now available. Paint specifiers are not too concerned with the composition or drying rate of solvents because their main interest is the dry film.

The paint user needs to be aware that the type of solvent used has a significant influence on the whole operation. The rate at which paint dries is a function of the solvent and some paints contain solvents which, if improperly handled, may be a health hazard. Fast-drying paints have solvents with low flash-points and these need to be contained and used in flame-free areas. Reputable manufacturers give clear health and safety information and guidance on the use of all their products so that users who follow this advice are unlikely to expose their operators or property to any risk.

Zinc in corrosion control (see also Chapter 5) The use of zinc metal coatings in the form of hot-dip galvanizing is very well established. In the code of practice BS 5493: 1977, metal coatings, mainly zinc, are recommended for use wherever a very long time to first maintenance is required. The Zinc Development Association quotes a life of up to 50 years for galvanized steel under some conditions.

Since zinc is anodic to steel it protects by sacrificing itself, even protecting areas where the coating is damaged such as at cut edges.

Calcium plumbate primer Among the surface treatments recommended in the current literature is the use of calcium plumbate primer. *This material should be avoided* because the adhesion of the paint can be uncertain and calcium plumbate contains a high proportion of lead.

Wash primers   These are two-pack paints containing polyvinyl butyral resin, zinc chromate and phosphoric acid. They are useful for coating small areas and providing that they are not applied too heavily and the surface is clean, they give good adhesion. On large areas they may cover contamination by grease and oil and contractors have been known to use them as a substitute for cleaning. Such a practice is almost certain to lead to premature adhesion failure and is not recommended.

Mordant or British Rail 'T' wash   The most satisfactory treatment for pretreating galvanized surfaces for painting is the use of Mordant or British Rail 'T' wash. This is a solution of copper salts in phosphoric acid and organic solvent. It should be applied to a galvanized surface which has already been cleaned and slightly roughened. It turns a steel surface black and should be rinsed with freshwater after about 15 min to leave a permanent black stain. If the stain is removed during rinsing, it indicates that the surface is not clean and the cleaning process must be repeated. Once the surface is dry it should be coated with a non-saponifiable primer that will not react with zinc salts. Chlorinated rubber or epoxy primers are ideal for this purpose.

Metal-sprayed zinc coatings   Metal-sprayed zinc coatings have a completely different structure from those obtained by hot-dip galvanizing. They are applied by melting a zinc wire in an oxypropane flame and atomizing and projecting the metallic particles in a stream of compressed air on to a blast-cleaned steel surface. The coating is porous and the surface is rough. Porosity accounts for approximately 10% of the volume of the coating.

Sherardized coatings   Zinc coatings may be applied to small components such as nuts, bolts and washers by sherardizing. Sherardized coatings are applied by tumbling the cleaned parts in zinc dust at an elevated temperature, but below the melting point of zinc. As zinc coated bolts are usually painted on site, sherardizing may be preferable to galvanizing because the surface is matt and free from grease. Epoxy or chlorinated rubber primers may be applied directly without the necessity of using wash primers or mordant.

Zinc-rich primers   Zinc-rich primers are paints pigmented with zinc dust as the rust inhibitor. Their composition is covered by BS 4562: 1971. There are three formulations covered by this standard but type 3 in epoxy media is the most popular for new steel. By definition in this standard, zinc-rich paint must have a 62–66% pigment volume concentration and 95% by weight of the pigment must be zinc. This corresponds to over 90% by weight of zinc in the dry film.

The organic zinc-rich primers give excellent corrosion resistance, particularly at low dry-film thicknesses. They are applied at up to 75μm thickness, but due to the high pigmentation overthick coatings can exhibit mechanical weakness; excessive thicknesses have been known to delaminate. These primers are usually overcoated with epoxies or chlorinated rubber which are not adversely affected by the zinc salts which may form at the interface.

Zinc silicates are primers used mainly on offshore structures. The original inorganic silicates were based on alkali-metal silicates and zinc dust. When mixed, they could be applied to blast-cleaned steel to give a completely inorganic cement-like coating containing zinc metal. The alkali silicates have been largely replaced by organic silicates in which the zinc dust is mixed with a partially hydrolysed alkyl silicate. These cure by hydrolysis through interaction with atmospheric moisture. Once hydrolysis is complete, the coating is completely inorganic.

Zinc silicates have a very good reputation for corrosion resistance under severe conditions and they do not salt as readily as the organic zinc rich primers. *They are far from user-friendly products* in that they require a very good standard or blast cleaning, Sa.3 is desirable, and great care must be taken to control film thickness and the atmospheric conditions during drying and curing. Low temperatures and low humidity will retard the cure rate. Overcoating may be a problem and a tie coat may be required. Areas of low film thickness are difficult to overcoat with additional zinc silicate.

A compromise has been reached with the introduction of organic modified zinc silicates such as PROTECTON 402. These are more user-friendly than either zinc rich epoxies or the inorganic silicates. They are tolerant to imperfectly blasted surfaces and may be applied to a specified minimum dry-film thickness of 75μm without risk of mechanical weakness on overthick areas. Furthermore, they can be overcoated with most paints after a very short time.

Heavy-duty coatings The materials and specifications described can be used for most heavy-duty applications provided that higher dry-film thicknesses are used. Bridges have been coated with epoxy systems applied at a fabrication works followed by chlorinated rubber or silicone alkyd on site. Chlorinated rubber is particularly useful for repainting weathered steel and either chlorinated rubber primers or water tolerant epoxy primers may be used for this type of site maintenance.

14.2 Paints or preservatives for absorbent surfaces

14.2.1 Timber (see also Chapter 19)

Timber used in construction can last indefinitely under favourable conditions. However, in many situations it is subject to hazards which may cause deterioration unless the species is naturally resistant or has been adequately protected. When considering the preservation and coating of timber, it is necessary to consider where it will be used, the hazards to which it will be exposed, and the natural durability of the timber species.

The principal causes of deterioration of timber in service are: (i) fungi; (ii) insects; (iii) marine borers; and (iv) physical and chemical attack.

Fungal attack   Fungal attack on timber normally occurs when its moisture content exceeds about 20%. The two main groups of fungi affecting timber are 'staining fungi' and 'wood-decaying fungi'.

Insect attack   Most timbers are susceptible to attack by wood-destroying insects, of which there are two main groups; (i) termites (*Isoptera*); and (ii) beetles (*Coleoptera*).

Marine borers   Timber in the sea or brackish water is liable to damage from marine borers. Very few timber species are resistant to attack. There are two groups of marine borers: molluscs and crustaceans.

Molluscs, which include shipworms, can grow to approximately 25 mm in diameter and 3 m in length, although the toredo found in temperate waters is smaller. The mollusc bores deep below the surface, producing tunnels that increase in size as it grows.

Crustaceans are small creatures, approximately 1–3 mm in diameter, and occur in intertidal waters. They tunnel close to the timber surface forming a spongy mass which becomes physically eroded. Crustaceans include *Limnoria*, commonly called the Gribble, which is found around the UK and in temperate waters.

Weathering   Timber exposed to the weather undergoes a number of chemical and physical changes on the surface due to the combined effects of moisture, temperature and sunlight.

Sunlight causes photodegradation of the lignin, which binds together the wood fibres, causing the formation of a grey unstructured mat of fibres on the surface. On some lighter timbers this is preceded by darkening.

This is often followed by the formation of surface cracks, or 'checks', resulting from rapid changes in the surface moisture content of the timber. The effect is most pronounced where variations in atmospheric moisture are accompanied by sudden and extreme changes in surface temperature.

Whilst the effects of weathering are mainly superficial, the break-up of the surface makes the timber more susceptible to infestation by moulds and other fungi.

Chemical effects   Chemical attack on timber may be caused by exposure to liquids, chemical fumes or air-borne pollution. Attack is normally more rapid at higher temperatures, and hardwoods tend to be more susceptible than softwoods. The use of timber in a chemically hostile environment must be fully evaluated. In some situations specialist coatings can be used to protect the timber against chemical attack.

Whilst most timbers are resistant to weak acids, wood is degraded by strong acids, oxidizing agents and alkalis. Solutions of salts cause differing effects, depending on the salt, temperature, duration of exposure and wood species. In some cases an increase in strength has been observed, in other cases a decrease. Exposure to steam, such as in high temperature steaming processes, can cause serious reduction in the strength of the timber.

Natural durability of timber   The natural 'durability' of a timber species is a measure of the resistance of the heartwood to fungal decay; the sapwood is usually susceptible. Durability may also be influenced by conditions and rate of growth, position of timber in the tree and storage and processing methods. The natural-durability classifications such as those used in the UK and the USA, based on the average life of heartwood stakes in ground contact, are given in *Table 14.1*.

Table 14.1   *Natural-durability classifications of some timbers[a]*

Classification	Average life of stakes (years)	Examples
Very durable	>25	Teak, greenheart
Durable	15–25	European oak, Western red cedar, Californian redwood
Moderately durable	10–15	Douglas fir, sapele, African mahogany
Non-durable	5–10	Eurpean redwood, spruce
Perishable	0–5	Beech, ramin

[a]Based on the average life of heartwood stakes in ground contact.

14.2.1 Types of wood preservative   Wood preservatives may be divided into three types: (i) tar oil; (ii) water-borne; and (iii) organic solvent. A further group includes specialist products for remedial treatment.

Tar-oil preservatives   The most common wood preservative in this category is creosote, produced by the distillation of coal tar. Different grades are obtained, varying in viscosity and volatility. All are highly toxic to fungi, insects and marine borers.

Water-borne preservatives   Water-borne wood preservatives are solutions of a single salt or a mixture of salts in water. They may be divided into two categories: (i) fixed-salt treatments; and (ii) water-soluble (leachable) treatments.

(i) Fixed-salt treatments Fixed-salt type preservatives contain a mixture of salts which react in the timber to become insoluble and highly resistant to leaching. They are normally applied by high-pressure processes, after which the timber must be redried. The overall process may result in dimensional changes in the timber. A variety of fixed-salt preservatives are in current use. The most common are the copper/chrome/arsenate preservatives.

(ii) Water-soluble (leachable) treatments The majority of water-soluble preservatives are solutions of boron compounds, or borates, in water. These preservatives are applied by the diffusion process to newly felled timber. Preservatives based on boron compounds are fungicidal and insecticidal, but their use is restricted to dry areas where leaching is not expected, such as in internal structural timbers.

Organic-solvent preservatives Organic-solvent preservatives contain fungicides and/or insecticides dissolved in an organic solvent, usually a light hydrocarbon solvent similar to white spirit. Most are highly resistant to leaching. Fungicides used include: pentachlorophenol (PCP), pentachlorophenol laurate (PCPL), zinc and copper naphthenates and acypetacs zinc. Insecticides include: lindane (γ-HCH) and permethrin.

14.2.1.2 Coatings for timber Coatings provide protection against weather, moisture, dirt, surface damage and provide a decorative finish. They may be broadly classified as follows:

- paints (opaque 'high-build' coatings)
- exterior wood stains (semi-transparent, coloured 'natural' finishes)
- varnishes (clear or tinted 'high-build' coatings)
- specialist coatings (high resistance to abrasion, chemical attack, etc.).

Typical uses of exterior coatings and maintenance periods are given in *Table 14.2*.

14.2.2 Absorbent surfaces other than timber

Mineral-based floors and walls can be successfully painted provided they are in sound condition and that an appropriate paint is selected. A very wide range of paint and coating systems are available. Each product is somewhat different and its detailed formulation is usually known only to the manufacturer. The successful use of sealers and paints on external walls has been reported, but caution is necessary in the specification and execution of these treatments. The alternative of a Portland-cement render may be viable in certain situations.

There are many good quality, durable structures and buildings where unfortunate weathering patterns have resulted in a

Table 14.2 Exterior coatings—typical applications

Use/timber	Alkyd gloss paint	Solvent-borne exterior wood stains			Water-borne	
		Low build	Medium build	Opaque	Acrylic gloss paint	Acrylic matt paint low build
Joinery	*				*	*
Cladding (smooth)	*	*	*	*	*	*
Shingles/shakes		*	*	*		*
Decking (smooth)		*	*	*	*	*
Fencing (smooth)	*	*	*	*	*	*
Plywood	*	*	*	*	*	*
Plywood (with overlay)						*
Sawn/rough[a]		*		*		*
Typical maintenance (years)	3–5	2–3	3–4	3–4	≥5	4–5

[a]Maintenance periods on rough sawn timbers may be 50% longer.

need for cleaning or painting. Dirt will not usually harm good quality concrete, so visually acceptable patterns of ageing and soiling do not need to be disturbed. There is great interest in the use of colours for whole or parts of buildings.

Waterproofing of walls and roofs is perhaps the most obvious application for paints and sealers as performance improvers. Wear-resistance properties can also be enhanced with certain floor sealers. Tough, strong paints, such as epoxy paints, should not be applied to relatively weak surfaces, otherwise fracture of the substrate may result.

Painting of mineral surfaces may be recommended for added durability, but the basis for this advice needs to be assessed carefully. The microporous and generally permeable nature of natural stone, plaster, cement render, concrete, etc., will be radically changed by the application of a surface barrier and certain paint films may lead to a build-up of moisture, that could, for example, result in frost damage in outdoor exposure. Partly because of this possibility, surface coatings are not generally permitted on highway structures.

Uncoated concrete should, theoretically, require little maintenance when made to recognized standards but some blemishes and poorer quality areas are inevitably produced. Anticarbonation, antichloride and moisture-ingress-resisting coatings are now more frequently specified for structural reinforced concrete in an attempt to improve durability, either for new structures or after repair is required.

BS 6150, ACI 515 and *BRE Digest 197* list the broad classes of paints available and their properties. They may be broadly categorized as:

- water-borne paints (lime washes; cement paints; emulsion paints; masonry paints; textured paints; multi-coloured paints; bitumen emulsions)
- oil-based paints (alkyd paints; urethane oil paints; epoxy ester and alkyd paints; phenolic paints)
- solvent paints (acrylic paints; chlorinated rubber paints; vinyl paints; fluoro-elastomer coatings; coal-tar paints; bitumen paints)
- two-pack systems (epoxy resin; polyester resin coatings; neoprene rubber; chlorosulphonated polyethylene; polysulphide rubber; silicone rubber; urea-formaldehyde; polyurethanes)
- moisture-curing polyurethane paints
- hot applied coating systems (coal tar; pitch; asphalt)
- clear finishes, stains and sealers (silicones; polymethyl methacrylate polyurethane resins)
- impregnants (based on silicones; silicates; acrylics; silanes; linseed oil; stearates; silicofluorides).

14.3 Intumescent coatings

Intumescence is a process in which, under the influence of heat, a solid substance transforms into an expanded, relatively rigid

Table 14.3 Comparative properties of various types of fire protection applied to steel structures

Protection type	Concrete encasement	Cladding panels	Cement sprays	Intumescent coatings
Installation speed	Very slow	Slow	Moderate	Fast
Interference with other trades	Great	Moderate	High	Negligible
Durability internal	Good	Good	Good	Moderate to good
Durability external	Good	Poor	Poor	Poor to good
Cost	High	Moderate to high	Low	Moderate
Appearance	Moderate	Good	Poor	Good
Conformability	Moderate	Poor	Good	Very good
Section range	Wide	Wide	Limited	Limited
Fire rating range	Wide	Wide	Moderate	Narrow

foam. This foam, having a lower thermal conductivity than the original substance because of the expansion, is used as a means of providing fire protection.

Commercial intumescent products are invariably applied to the surface materials of a construction as a means of imparting resistance of the material to fire, or protecting the structure from the effects of fire.

The interest in the use of intumescent coatings as a means of structural fire protection is largely derived from their apparent ease of application and other properties, when compared with other fire protection systems. The properties of various types of fire protection listed in *Table 14.3* illustrates this. Thus, it can be seen that, apart from the durability aspect, intumescent coatings have strong advantages.

A sample formulation for an intumescent coating is shown in *Table 14.4*. The performance of the material may be varied by changes in formulation. Specific requirements can be met by discussion with specialist manufacturers and verified, if necessary by test. The formulation in the table is stated to obtain a 1 h rating on the standard 170 HP/A beam with about 1.8 mm dry film thickness to BS 476 conditions.

The mode of action of an intumescent product under the influence of heat can be simplified into the following sequence, as the temperature increases.

(1) The binder system softens to a highly viscous melt. Generally this occurs from 70°C.
(2) Some of the plasticizer and spumescent decomposes, emitting inert gas that blankets the surface of the coating, preventing inhibition of the organic components by the fire source. This will be observed from 90°C up to and beyond the point of intumescence.
(3) The catalyst loses ammonia, releasing liquid acid and lowering the viscosity of the melt. This occurs progressively from 150°C upwards.

Table 14.4 *Sample formulation*

Constituent	Function	Content (%)
Pliolite VTAC-L resin 70%	Binder	10.0
Solid chlorinated paraffin	Plasticizer	8.0
Exolit 422 ammonium polyphosphate	Catalyst	25.0
Dipentaerythritol	Carbonific	7.5
Melamine	Spumescent	7.5
Tiona 535 rutile titania	Pigment	9.0
PRS	Solvent	32.6
Bentone 38	Thixotrope	0.3
Methanol	Solvent	0.1
Total		100

(4) The carbonific melts and reacts with the liquid acid, reducing to carbon and releasing water vapour.
(5) The spumescent decomposes releasing inert gas that expands the melt into a soft foam. Reactions (2)-(4) will occur simultaneously at about 210°C.
(6) The foam partially decomposes and sets into a relatively rigid char.
(7) The residual acid reacts with the pigmentary materials, if present, to form refractory phosphates.
(8) The carbonaceous char is gradually ablated by the fire source reducing the strength of the char.

This explanation is an oversimplification. Many side-effects occur and, generally, all these stages, and others, occur at the same time. In addition, the course of the reactions can be mediated and altered by the addition of a variety of functional agents.

Bibliography

DORAN, D.K. (ed.), *Construction Materials Reference Book*, Butterworth–Heinemann, Oxford (1992):
  Chapter 31: Cunningham, D.E.J.
  Chapter 32: Crookes, J.V.
  Chapter 33: Lawrence, C.D.
  Chapter 34: Aslin, D.C.

15 *Plaster and plasterboards*

Historically, lime and sand was the main plastering medium used in the UK, with either cement or gypsum added as a binding agent from the middle of the nineteenth century. Since the end of the last war in 1945, however, when the vast rebuilding programme demanded speedy construction methods, gypsum plasters have increasingly dominated the scene.

Gypsum plasters, which, being premixed, require only the addition of clean water on site, are light in weight, possess a controlled chemical set and offer excellent adhesion to a wide variety of backgrounds, have enabled the craft to keep pace with other major changes in the industry. The controlled set provides the plasterer with continuity of production, thereby eliminating the costly waiting time between coats necessary with sanded or cement-based plasters.

Other benefits that attract specifier and user alike include good fire resistance properties, use in low temperature conditions where plastering with sanded undercoats would not be possible, clean sites, easy estimating and ordering and a reduction in errors of mixing and proportioning. Gypsum plasters also provide a perfect base for subsequent decoration.

The use of gypsum to support decorative finishes is not a new idea. There is evidence that the mineral was used some 3000 years ago by the Egyptians who used it within the Pyramids. In the UK, the basic raw gypsum material is mined at seven sites across the length and breadth of the country from Robertsbridge in East Sussex to Kirkby Thore in Cumbria by British Gypsum Ltd, a subsidiary of BPB Industries plc. In addition to being the basic component in British Gypsum's wide range of bagged plasters, gypsum lies also at the heart of an extensive series of plasterboard products and systems available from this company in the UK and from other BPB Industries' subsidiaries throughout mainland Europe, and in Canada and India.

15.1 Plaster range

Carlite premixed plasters Carlite is a range of lightweight, retarded hemi-hydrate premixed gypsum plasters that require only the addition of clean water to prepare them for use. The range includes four undercoat grades and a single finishing product.

Carlite Browning is an undercoat for solid backgrounds of low to moderate suction and an adequate mechanical key; Carlite Browning HSB is required for solid backgrounds of a very high suction. Carlite metal lathing is an undercoat that has been specially developed for expanded metal lathing and Carlite bonding is an undercoat for low suction backgrounds, soffits of composite floors with concrete beams, plasterboard, expanded polystyrene and surfaces treated with a polyvinyl acetate co-polymer (PVAC) bonding agent. Carlite finishing is a final coat for use with all of the above listed Carlite undercoats.

Whilst dry, bagged plaster is not affected by low temperatures; plastering on frozen backgrounds must be avoided and the use of such products is negated where the temperature exceeds 43°C.

Thistle board finish plaster   Thistle board finish is a final coat for grey-faced plasterboard and normal ballast concrete suitable for plastering direct. It can be applied for plastering direct. It can be applied in two coats at a total thickness of 5 mm or in a single coat at 3 mm total thickness. When applied as a single coat the plasterboard joints must be reinforced with either Gyproc joint tape or glass fibre, self-adhesive tape.

Thistle universal one-coat plaster   This is a gypsum-based plaster suitable for hand application to most types of internal background. It may be used in one coat at thicknesses suitable for all normal purposes on brick, block, concrete, plasterboard and expanded metal lath. Of white finish, this plaster is a hemihydrate with special additives that improve the characteristics of workability and application.

The recommended thickness of application is 13 mm for walls other than concrete, where it should not normally exceed 10 mm.

Thistle renovating plasters   The renovating range of plasters comprises an undercoat and a finish coat especially designed for use in older properties where moisture is still retained after successful damp-proof-course treatment. The undercoat provides early strength in damp conditions and speedy surface drying, whilst the finish has good working and setting characteristics together with high strength—although it is essential to use Thistle renovating finish over Thistle renovating plaster as soon as the undercoat has set.

It is important to note that these plasters should not be used directly on backgrounds in damp areas below ground level, or with damp-proof-course renewal systems involving either electro-osmosis or cementitious grouts.

Thistle renovating plaster is a premixed lightweight gypsum undercoat plaster containing special additives; it should be applied in thicknesses of 11 mm to cover an area of approximately $120 \, m^2$ per 1000 kg.

Thistle projection plaster   Thistle projection plaster has been formulated for single-coat application by machines that mix and spray in one operation. The plaster is levelled and trowelled in the conventional way and supplied in a single multipurpose grade for general internal use.

This plaster product consists of gypsum hemi-hydrate plasters combined with a number of additives to increase water retention, improve plasticity and stabilize the setting time. It should be applied in a thickness of 13 mm to walls other than concrete; on concrete soffits, the thickness should be 5–10 mm, and if the background is level and of an even suction, for instance, plasterboard, it may be possible to reduce the overall thickness to just 5 mm.

Thistle X-ray plasters   Thistle X-ray plaster is a specially formulated undercoat which is applied in the traditional manner to provide standards of protection from electromagnetic radiation as required by the National Radiological Protection Board.

It is used in situations where X-ray generating equipment is installed, to achieve the required degree of radiation protection, in such areas as hospitals and medical research laboratories.

A final coat of Thistle multi-finish is required to provide a high standard of conventional, internal finish, but it is important to note that the Thistle multi-finish does not contribute to radiation protection, which is provided entirely by the specified thickness of Thistle X-ray undercoat.

Thistle hardwall plaster   Thistle hardwall plaster is a premixed, undercoat plaster containing a blend of special aggregates and additives which combine to provide numerous benefits to both specifier and user. In addition to gypsum plaster's well known characteristics such as controlled set, continuity of production, freedom from inherent shrinkage cracks, excellent adhesion and ease of working, Thistle hardwall also provides a quicker drying surface and stronger resistance to efflorescence compared to other types of gypsum plaster. It also provides superior impact resistance which is useful in areas of high usage. Thistle hardwall is suitable for use with a wide spectrum of brick and block substrates, expanded metal lath and application by plaster projection machine as well as by the manual method. When applied by machine the plaster is worked as an undercoat plaster in the traditional two coat plastering system. The required final coat plaster is Thistle multi-finish which is applied by hand.

Thistle multi-finish plaster   Thistle multi-finish is a new generation final coat plaster and has been specially formulated to perform equally well on a variety of backgrounds including grey and decorative faced Gyproc plasterboards, Thistle hardwall, sanded undercoats and normal ballast concrete suitable to receive a thin application of plaster.

When used on undercoats of Portland cement and sand or Portland cement, lime and sand, it is important to allow the undercoat to mature before applying the finishing coat. Failure to do so could result in shrinkage cracks appearing on the

surface or a complete loss of adhesion of the finishing coat from
the undercoat.

15.2 Plasterboard

Plasterboard is the name given to building board made of a core
of aerated gypsum plaster usually enclosed between two sheets
of strong paper.

Plasterboard dry linings continue to gain rapid popularity as
an alternative to plaster finishes for a variety of important
reasons. They are widely used (independently or as part of a
system) in new construction and rehabilitation work to meet
not only internal lining requirements but also to provide protection against the hazard of fire and condensation, or to
enhance thermal or sound insulation.

In addition, the use of plasterboard dry lining systems means
that little water is introduced into a building's structure at a late
stage, which reduces the drying out time required. This gives
greater freedom in the selection of the type of first decoration.

On site, dry lining operations are less arduous than wet
plastering, a minimum of cleaning up is required and, as the
services can be accommodated in the cavity between plasterboard and wall, chasing is also reduced substantially. Add to
this careful consideration at the design stage, proper site orga-
nization and workmanship, and plasterboard dry lining can be
favourably comparable in cost with good quality two-coat
plastering.

When correctly fitted and jointed, the grades of plasterboard
that are available with tapered edges will produce a surface that
is as smooth as traditional plastering but with the added benefit
of easier maintenance as cracking is virtually eliminated.

Over 80% of the gypsym plasterboard and plasterboard systems sold in the UK are sold under the brand names Gyproc,
Thistle, and Glasroc. However, the plasterboard manufactured
for dry lining is different from the grades produced to receive
wet plasterwork, and this is identified by an ivory coloured
surface onto which internal decoration may be directly applied.

15.2.1 Plasterboard range
Gyproc wallboard This board has an ivory coloured exposed
surface and a grey surface on the back; its edges can be supplied
tapered or square, depending on the joint treatment required. It
is also available in Duplex grade which has a backing of metal-
lized polyester film. The film acts as a reflective thermal insulator when the board is fixed facing a 20 mm minimum air
space, and as a vapour control layer which reduces the passage
of water vapour from inside the building into the structure and,
therefore, helps to reduce the risk of interstitial condensation.

Gyproc fireline board This is similar to Gyproc wallboard but
is only available in 12.5 and 15 mm thicknesses, with tapered

edges. The gypsum core of this board incorporates glass fibre and additives to improve its fire protection performance.

Gyproc plank   Gyproc plank is also similar to gypsum wallboard, but thicker and normally supplied with tapered edges for dry lining purposes. A square edge plank with grey surface on both sides is used as a first layer in certain two-layer lining systems. Gyproc plank is also available in Duplex grade.

Gyproc Thermal Board Plus   This is a plasterboard thermal laminate with a backing of extruded polystyrene. It is available in six thicknesses, ranging upwards in 5 mm increments between 25 and 50 mm thicknesses. The most recent addition to the Gyproc range of thermal laminate boards, Gyproc Thermal Board Plus, has been developed to meet the April 1990 Building Regulations in respect of the new 0.45 W m^{-2} K external wall U values.

Gyproc thermal board   This is a laminate composed of gypsum wallboard bonded to a backing of expanded polystyrene to provide a good standard of thermal insulation and is also available in a vapour-check grade, which incorporates a vapour control layer at the interface between the wallboard and the expanded polystyrene. This membrane reduces the passage of water vapour from inside the building into the structure and, therefore, can help to reduce the risk of interstitial condensation.

Paramount panels   A prefabricated gypsum wallboard panel formed from two wallboards (with exposed ivory coloured faces), these panels are separated by, and bonded to, a core of cellular construction. They are available in 57 and 63 mm thicknesses, with either tapered or square edges. It is used primarily as a lightweight partition in housing.

Gyproc cornice range   Although the ceiling/wall angle in a room may be finished in the same way as with any other joint, plasterboard cove provides a very attractive finish. In addition to applications in new dwellings, coving is becoming increasingly popular in the do-it-yourself and general home redecoration marketplace.

15.2.2 Performance

Fire protection   Due to the unique behaviour of the non-combustible gypsum core, plasterboard linings provide good fire protection when subjected to high temperatures.

Plasterboard is combustible (when tested to BS 476: Part 4: 1970, although in the *Building Regulations 1985*, plasterboard is designated a 'material of limited combustibility'), the ivory (exposed) and grey (backing) surfaces of standard Gyproc wallboard, Fireline board, plank and Duplex grade surfaces are class 0 when tested to BS 476 Parts 6 and 7 and in accordance with the requirements of the Building Regulations. Class 0 is the highest standard required for restricting the spread of flame over wall and ceiling surfaces.

Periods of fire resistance for a few constructions that incorporate gypsum plasterboard dry linings are given in Approved Document B2/3/4.

Sound insulation For maximum sound insulation to be obtained for a building element on site, all air paths such as perimeter cracks and gaps should be sealed. This can be achieved, for example, by the use of acoustical sealants around perimeters, or gypsum cove at the wall and ceiling angles. Ideally, a building element should be imperforate for optimum sound insulation.

Thermal insulation Due to its aerated gypsum core, plasterboard is a low-thermal-capacity lining material. Plasterboard dry-lining systems can improve the thermal insulation of a building element and, therefore, reduce heat loss. When used in intermittently heated buildings, a plasterboard dry lining responds to heating more quickly than denser plaster finishes and allows rooms to warm up more rapidly.

Effect of high and low temperature Gypsum plasterboards are unsuitable for use in temperatures above 49°C. They can be subjected to freezing temperatures without risk of damage.

Condensation To prevent condensation occurring, adequate heating and ventilation are necessary in addition to good thermal insulation. The absence of any of these requirements can result in condensation and mould growth on internal surfaces (surface condensation) or potentially harmful condensation within the structure (interstitial condensation).

Bibliography

O'SULLIVAN, D., In DORAN, D.K., (ed.), *Construction Materials Reference Book*, Butterworth–Heinemann, Oxford, Chapter 35 (1992)

16 *Polymers*

16.1 Introduction

16.1.1 What are polymers?
First reactions to this topic might well be based upon an assumption that polymers are comparatively new materials. This is far from the case as one of the oldest construction materials—timber—is in fact polymer based. So too are animal glue and natural rubber.

Polymers fall into two main categories—those which after reheating may be repeatedly reshaped (thermoplastics) and those which cannot (thermosets).

16.2 Natural polymers—harnessed by man
The first of the man-modified materials was rubber. Coagulated natural rubber latex is of very little practical use because, although under certain conditions it is soft and elastic, it becomes very hard in cold conditions and soft and sticky in warmer weather. It was the development of vulcanization in 1844 which overcame these difficulties and gave us the rubber with which we are familiar in waterproof membranes, flexible seals, water-bars, bridge-bearings, electrical insulation, etc.

16.3 Synthetic polymers—thermosets
The first completely synthetic polymers were phenol–formaldehyde resins which were brought to commercial realization with the development of Bakelite in 1908. With its good electrical insulation properties and comparative ease of moulding and extrusion, this paved the way for the mass production of electrical components at a time when the electricity supply and distribution industry was developing and expanding rapidly. Although functionally very satisfactory, the colour range of phenolic moulding materials was limited to black, and dark shades of brown. The development of urea–formaldehyde resins in the early 1930s allowed the production of white and pastel shades. The physical properties of these materials were not as good as the phenolics, being somewhat sensitive to moisture and tending to discolour on heating, but they fulfilled a useful role until melamine–formaldehyde resins (1940) offered a superior, more stable, but more expensive product. These various products, still popularly known by the most familiar trade name 'Bakelite', are thermosetting materials.

16.4 Synthetic polymers—thermoplastics
The cellulose plastics were all thermoplastic, which made them much easier to process and fabricate into complex shapes and

allowed the development of injection moulding from technology of the metal die-casting industry. The range of thermoplastic materials widened considerably during the 1930s with the development of polymers synthesized from coal or oil feed stocks; polystyrene, acrylics and vinyl polymers. Polystyrene had no direct bearing upon the construction industry until the 1960s when expanded polystyrene emerged as a very efficient and economical thermal insulation material. It has also become accepted as a useful low-cost void former in concrete construction and a floor slab underlay to accommodate clay heave. Vinyl polymers in the form of plasticized PVC gradually replaced rubber for electrical insulation and the lead sheathing widely used at one time for the protection of elastic cables. The chemistry and technology for processing PVC without plasticizers was developed in the late 1940s and when the economics became right, rigid PVC (UPVC) played a major contribution in construction, largely replacing cast iron for rainwater goods, soil, drainage, and water supply pipes. Light weight, slight flexibility, ease of colouring and production in continuous lengths simplified some aspects of the industry. Being less brittle than phenolic materials, but having fair rigidity and good durability, rigid PVC opened up the possibilities of plastics forming some part of the fabric of buildings rather than just fittings and artifacts. Extruded UPVC sheeting began to compete with asbestos-cement and corrugated steel sheet for industrial roofing and wall cladding. Plasticized PVC was also used as a permanent coating on steel sheet cladding. Acrylic sheeting also entered this market when its scale of manufacture made it competitive and this offered the additional advantage of transparency. This, coupled with lower density and better resistance to breakage than glass, made acrylics ideal for many lighting applications. Where physical abuse might be encountered, clear polycarbonate resins provided the answer to glazing applications in the 1960s. UPVC has also become prominent in fenestration, as window frames. Being fabricated from multi-wall hollow extrusions, UPVC frames contribute significantly to thermal insulation and sound attenuation and, in the replacement market at least, are competitive in price.

Opaque, pigmented acrylic sheeting and mouldings have complemented the plastics pipe revolution as they provide the basis for the new generation of sanitary ware. Baths, basins and shower trays are thermoformed from acrylic sheet and tap tops moulded in acrylics. Complete taps are moulded in polycarbonate.

The development of polythene also proved to have a tremendous impact upon the construction industry. Initially only used as a high dielectric insulator, when its price came down it became another contender for the insulation of wiring and fittings, but its major role in building and construction came in the form of extruded tubing and pipes now widely used in the gas and water supply distribution systems. It is also universally used as a damp-proof membrane.

16.5 Coatings, castings and laminates

Natural polymers have been used for centuries in surface coatings, but synthetic resins widened the range of binders or 'vehicles' for paint manufacture and improved their quality. The term 'polymers' is generally associated with the plastics industry, but the demarcation between paints and plastics is simply historical. Some reactive resin systems (e.g. epoxies) are applied as liquids without solvents and they harden to give coatings of greater build and better resistance to aggressive chemicals and abrasion. Some thermoplastics (and even some thermosetting resins) can be applied as hot-melt coatings with high build. All of the resins mentioned so far have been used in the coating industry, but alkyd resins were developed specially for this purpose and have been the basis of many of the gloss paints used in the building industry for several decades. Chemically these materials are polyesters, being reaction products of polybasic acids and polyhydric alcohols, such as phthalic acid and glycerol. This development in the 1930s led, later, to the development of polyester casting resins and then their use as low pressure laminating resins. This was the beginning of the glass reinforced plastics (GRP) industry and was to have a profound impact upon the construction world.

A very similar course of development was progressing at about the same time using epoxide resins. Both epoxy and polyester resins, when used as a binder for glass fibre fabrics (both woven and non-woven), produced sheet materials of considerable strength, chemical and weather resistance. Being amenable to low pressure processes, in contrast to the very high pressures required for phenolic and melamine laminates, these newer laminates could be made without heat, by a simple hand lay-up process, in comparatively cheap and simple moulds in almost any complex shape, whereas phenolic and melamine laminates produced in multi-daylight hydraulic presses were limited to flat sheet form.

The design freedom and economic production of the hand lay-up process made these GRP laminates ideal for chemical plants, process vessel linings etc, but also suggested their use in structural building application. Whilst they enjoy the high tensile strength of the glass fibres, their elastic modulus is only one-tenth that of aluminium. With material costs making very thick section prohibitively expensive, the most efficient way of achieving rigidity was by stressed skin or monocoque design. The resulting structures were of folded plate or curvilinear form, which rather limited their use to industrial or leisure buildings. Economic use was made of GRP laminates as sheet claddings in conjunction with steel or timber frames.

16.6 Resin mortars

When using these resins to produce castings, it was necessary and more economical to use a high loading of mineral fillers. This soon made apparent the possibility of using them as binders for sands and aggregates, to produce resin mortars or concretes having much of the versatility of hydraulic cement concretes, with greatly enhanced chemical resistance, abrasion resistance and flexural strength. These resin mortars are widely used for the manufacture of artefacts such as chemically resistant process plant, sanitary ware, drainage gulleys, large diameter sewer pipes, etc. They also find considerable use as functional and decorative floor screeds and concrete repair materials.

16.7 Polymer emulsions

Reference has already been made to the use of polymers in solution for surface coating and impregnation. Some polymers can also be dispersed in non-solvents; for instance water-dispersed epoxy resin systems are used as floor coatings. This is analogous to the way in which oil-in-water or water-in-oil emulsions can be prepared for such applications as machine cutting oils. The same technique of mixing two incompatible liquids with a trace of surface-active agent until one is broken down to fine stable droplets uniformly dispersed in the other, is also used in the preparation of some polymers from monomers. The monomers, which may be volatile oily liquids or gases under pressure, are dispersed in water and reacted with suitable initiators or catalysts, until the suspended monomer droplets become suspended particles of polymer. The end product is often referred to as a polymer emulsion (but polymer dispersion is the correct term) and is very similar in many ways to natural rubber latex. Many synthetic rubbers can be prepared in this manner and one, styrene–butadiene rubber, is particularly used in the construction industry. Many vinyl and acrylic polymers and co-polymers are also produced as water-based dispersions for construction work. Co-polymers are polymerized from blends of different monomers to combine the advantage of two or more basic products.

When spread out in a thin layer, most of these polymer dispersions lose their water by evaporation and the polymer particles coalesce together to form a coherent film. This has applications for surface coating and many vinyl co-polymers are the bases of emulsion paints. Back in the 1920s it was found that natural rubber latex, if stabilized to resist coagulation in an alkaline medium, could be mixed with Portland cement to produce a tough flexible coating or for use as an adhesive. Many synthetic polymer dispersions can be similarly used and find particular application in upgrading the strength and abrasion resistance of floor screeds and as bonding agents between fresh wet concrete and existing hardened concrete. When incorporated into cementitious mortars they can substantially reduce the permeability of the mortar to liquids and gases, which accounts for their considerable use in concrete repair compositions.

16.8 Seals and sealants

Low modulus polymers have become essential to the construction industry in the formation of flexible seals at joints between structural elements. Preformed water-bars and joint seals, often of quite complex cross-section, are extruded from plasticized PVC, natural or synthetic rubbers. Other joint seals are formed *in situ* by the pouring of hot materials—originally bitumen, but much better performance and durability can be achieved with rubber–bitumen blends or pitch–PVC. Hot pouring is of course limited to horizontal joints, but thixotropic cold-curing polymers such as polysulphide rubbers, polyurethanes and silicones can be gun-applied into vertical joints.

16.9 Inorganic polymers

These materials represent a contrast to other polymers so far discussed, in that hitherto we have been considering wholly organic polymers, in which the repeating units always had carbon atoms linked together as their backbone. Silicon has certain similarities with carbon in its atomic structure (each has four electrons in its outer shell) and this is reflected in some analogous compounds. One result of this is the ability to form polymer chains in which alternate silicon and oxygen molecules form the backbone and this chemical linkage is more stable than that between carbon atoms, so that these polymers can withstand much higher temperatures and are more resistant to many aggressive chemicals. The generic term for polymers based upon a silicon–oxygen backbone, often with organic branches, is silicone. Silicone polymers come in many forms ranging from oils to rubbers, to hard resin coatings.

Silicone polymers have very good water-repelling properties and this is put to good use by reacting them inside the pores of concrete to form a hydrophobic lining, which prevents the capillary movement of water without blocking or sealing the pores. Concrete thus treated is resistant to the passage of liquid water, but water vapour passes freely allowing the concrete to dry out if necessary. This is achieved by impregnating the surface of the concrete with a silane monomer or a siloxane oligomer (a very short chain silicone–oxygen species which is less volatile than the silane monomer). These materials are of such small molecular size that they can easily penetrate the finest pores in concrete and then react with free lime or moisture to polymerize and form a silicone–resin lining.

16.10 Reactive polymers

Many of the polymers considered are converted from liquid or semi-liquid materials into solids or rubbery solids after having been fashioned into the shape required. This means that part of the polymerization reaction is carried out on site and there are two types of chemical reaction employed. In 'addition' reactions, a chemically reactive polymer such as epoxy resin is mixed with another reactive chemical (such as an amine hardener or curing agent) which is then able to form chemical links between reactive sites on the polymer chains and connect them together into a fairly rigid network. By contrast, polyester resins contain the two reactive components (a long chain ester polymer and styrene monomer) but these are prevented from reacting with each other by the inclusion of an inhibitor. When the reaction is required to proceed, a very small quantity of a catalyst chemical (usually a peroxide) is added. For ease and safety of handling, the peroxide catalyst is usually supplied dispersed into a large volume of inert mineral filler. The use of reactive chemicals on site does require disciplined procedures and these are clearly stated in the health and safety data sheets issued by the material suppliers.

16.11 Rubbers

A number of elastomeric polymers have been developed since the late 1930s, initially to avoid dependence upon the Far East for rubber feedstocks and to circumvent the variability found in natural rubber in those days. Styrene–butadiene rubbers, butyls, polychloroprene, polysulphides, polyurethanes and silicones, to name the principal groups, have all been developed as engineering materials. Each has some particular characteristic, either in mechanical properties or chemical resistance. Meanwhile the production of natural rubber has been brought under control to provide reliable materials in a range of grades, with some performance properties which still cannot be matched by the synthetics.

The construction and building industry uses these rubbers for flexible seals and gaskets, water-bars, bridge bearings, vibration attenuating mountings, membranes for roofing and the lining of reservoirs etc. Flexible hoses carrying compressed air and water around construction sites are also dependent upon rubber.

16.12 Engineering properties

The arrival of resin-based high strength materials presented a challenge to mechanical and structural engineers. All their design experience had been founded upon generations of use of metals, timber, stone and concrete. Resin binders presented them with new concepts which many have still not mastered. Perhaps the greatest difficulty encountered has been coming to terms with the effect of heat.

The physical and mechanical properties of traditional materials, steel, concrete, masonry, etc. are comparatively constant over the normal range of working temperatures, say $-10°C$ to $+30°C$. Although polyester and epoxy resins are technically thermosetting materials, their properties may vary considerably over this same temperature range. Their elastic modulus reduces with increasing temperature, but the effect of this change can be minimized by chemical tailoring of the resin to give a high degree of cross-linking and by using the maximum filler loading compatible with maintaining adequate workability.

16.12.1 Fire All organic materials are combustible, but the ease of ignition and rate of combustion depend upon many factors. The first is the ease with which the bonds holding the atoms together may be broken. This varies considerably with molecular configuration. Thus polyethylene, which may be regarded almost as an ultra-high molecular weight paraffin wax, ignites easily and will support combustion, i.e. will continue to burn when the source of heat is removed. Phenolic resins, which also consist of carbon and hydrogen atoms, but linked together in a very different pattern and have an occasional oxygen atom included, will not support combustion and will only burn whilst held in a strong flame.

The inclusion of other atoms in the molecule can have a considerable effect upon combustion. Thus, PVC, which empirically may be considered as polythene, is also very reluctant to burn. The flame retardant benefit of chlorine and other halogen atoms can also be achieved by blending a halogen-rich chemical into an otherwise flammable material. The effect can be greatly enhanced by including antimony oxide in the mix. The incorporation of large quantities of inert mineral fillers can have a substantial effect, particularly fillers which lose water of crystallization when heated. The ratio of mass to surface area also plays a part in regulating rate of combustion.

16.12.2 Weathering The molecular bonds of polymers may also be vulnerable to rupture by the effect of ultraviolet radiation. Again, some polymers are much more stable than others and the stability of all polymers can be greatly enhanced by compounding with suitable chemicals. Heat and oxygen can also encourage molecular breakdown, but these effects can be inhibited by the incorporation of anti-oxidants and other stabilizing chemicals. PVC, which has comparatively poor resistance in its unmodified state, when suitably compounded has been shown to give excellent durability in the form of rainwater goods, vinyl coated steel cladding, etc. Acrylic resins, which are intrinsically more stable give quite outstanding performance in the field of surface coatings, where a life of fifteen years is now the expected norm.

16.13 Adhesives

Traditional glues may be classified under the headings: animal, casein (or other proteins), vegetable, mineral and organic. Apart from animal glues most are of low strength and have largely fallen into disuse. Since animal glues are technically polymers it follows that most adhesives used in construction are polymer based. These include epoxies, two-polymer adhesives (where the properties of one polymer are modified by the addition of another), phenol-formaldehyde and polyurethane.

Since the formulation of these products is being continuously updated by manufacturers it is impossible to give safe general guidance on properties. It is therefore suggested that information is sought from suppliers. Three important factors relating to the performance of adhesives are:

- failure stress (failure load divided by glued area)
- creep behaviour
- durability (including under fire).

Specifications for adhesives deal with these properties, prescribe test methods and state minimum acceptance values. Care must be taken to match the strength of an adhesive with the inherent strength of material(s) to be glued.

16.13.1 Surface preparation

Of cardinal importance in using an adhesive is the preparation of mating surfaces.

Typical surface preparation procedures for various substrates are described below.

Concrete   The surface of concrete and other cementitious surfaces should be carefully prepared, preferably by mechanical means, to give a clean lightly profiled sound surface. Typical methods include scabbling, needle gunning, grit blasting and high pressure water jetting. Acid etching, generally with 10% hydrochloric acid, can also be used but care must be taken to ensure that all traces of the acid, etc., are removed. In general, the surface strength of the prepared concrete should be greater than $1 \, \text{N} \, \text{mm}^{-2}$ in tension.

Steel   All steel surfaces should be degreased and abraded to a bright metal finish, ideally by grit blasting to SA 2.5, immediately before bonding. If immediate bonding is not possible, a solvent based epoxy resin holding primer should be applied. Excellent bond strengths are achieved to well prepared conventional mild steels. However, the bond strengths to some grades of stainless steel are less good and special complex chemical surface treatments may be needed to achieve optimum bond performance, although in most applications the bond is more than adequate if the surface is abraded well.

Brick   Brick surfaces should be clean and free from all friable material. In general, scrubbing with a hard bristle brush is adequate. Some bricks have very weak surfaces and should never be bonded with high strength resin adhesives, particularly in external applications.

The bonding of brick slips on to reinforced concrete as an architectural feature is often specified. The use of some form of mechanical support in conjunction with an adhesive is to be recommended. Soft joints at appropriate intervals to accommodate the differential movement between the bonded/pointed slips and the concrete substrate must be incorporated.

Wood Most woods simply require abrading with a medium abrasive. However, some woods may exude oily substances and require careful degreasing before bonding.

Glass Glass should be carefully cleaned with a warm detergent solution and dried carefully. The durability of the bond to glass can be markedly improved to pretreating the glass with a solution of a suitable silane coupling agent.

Rubber Different rubbers may require complex surface treatments to achieve optimum bond performance. Natural rubber, which is commonly used in the fabrication of rubber bridge bearing systems, often contains processing waxes which markedly affect the bond strengths. To ensure high bond strengths the surface should be carefully treated with an appropriate solvent to remove all traces of wax or grease and then carefully treated with concentrated sulphuric acid followed by washing with water and drying. The acid causes the surface of the rubber to harden and form microcracks which provide an excellent key for the adhesive.

Glass reinforced plastics Glass reinforced plastics simply require careful degreasing with an appropriate solvent and abrading to give a lightly profiled surface.

16.13.2 Applications
Interesting and recent applications include segmental construction of bridges using stress-distributing adhesives, strengthening of reinforced concrete structures by external bonding of steel plates, resin injection to repair cracked concrete structures and the development of an adhesive for skid-resistant road surfaces.

16.14 Geotextiles

16.14.1 Introduction
Geotextiles are polymer-based reinforcements often used within a soil structure. Like any other material, the fitness of a geotextile to serve a particular purpose is formalized in a specification of required properties. The procedure by which these required properties are evaluated is, in essence, the design process. The empirically derived specification may be limited in as much as experience may be limited to a specific type of geotextile used in a limited range of conditions. Although a specification derived from this type of approach may be difficult to extrapolate to other conditions, it has the advantage that it incorporates the geotextile properties necessary both to perform the required function and to survive the rigours of the installation process.

In contrast, a specification derived from an analytical technique can account for a wide range of parameters which might arise in a certain application. However, with this approach care must be taken to define the required geotextile properties both to fulfil the design function and to survive the installation process. The four cardinal functions credited to geotextiles are: (i) separations, (ii) filtration, (iii) drainage and (iv) reinforcement. For a particular application one or more of these functions will be of primary importance, as illustrated in *Table 16.1*.

The ranking of importance of these functions is very much conceptual because it arises from the perceived design function of the geotextile once installed. For the geotextile to perform the required design function, it must first survive the installation process without serious impairment. An example of this is the use of a geotextile filter to wrap a coarse granular drainage medium in a French drain. Here the prime design properties would be an appropriate geotextile pore size to retain soil particles in the base soil and an adequate flow capacity normal to the plane of the geotextile to transmit flow from the base soil to granular drainage medium. There is no requirement for geotextile strength at this level of assessment, however, such a requirement becomes self-evident when the construction process is considered. The geotextile would not be installed with the precision of a watchmaker and in all likelihood would be draped in the drainage trench and forced to comply with the profile of the trench by the weight of the granular drainage medium which would be placed by mechanical shovel or end-tipping. Clearly, if the geotextile did not have adequate resistance to puncture, tear and bursting, it would be unlikely to survive this process. Consequently, the ability to survive, or the 'function' of survival is of paramount importance and must be considered as part of the design.

Having defined the required geotextile properties for survival and performance of the design function, it remains to select a conforming product. This selection demands considerable thought and care because the properties of geotextiles vary widely according to geotextile structure, polymer type and the test methods employed to measure these properties. These aspects are considered in the following sections.

Geotextile fabrics generally fall into one of the two broad categories: woven or non-woven textiles. There are other fabric structures such as knitted fabrics or malimo stitch bonded fabrics. However, these are in a small minority and are, therefore, not considered further. As well as geotextile fabrics there are related non-textile products such as grids, meshes, mats and strips which may perform some of the geotextile functions. Most commonly used is polypropylene.

16.14.2 Mechanical properties

The most relevant mechanical properties for this material are:

- grab strength
- puncture strength
- burst strength
- trapezoidal tear.

Table 16.1 Relative importance of main functions in selected applications

Application	Separation	Filtration	Drainage-in-plane	Reinforcement
Trench drains	Secondary	Primary	Not relevant	Minor
Wall drainage	Not relevant	Primary	Primary	Not relevant
Embankments—basal drainage layer	Secondary	Primary	Not relevant to secondary	Not relevant
Subgrade stabilization	Primary	Secondary	Not relevant	Primary to secondary
Embankments—basal reinforcement	Primary to secondary	Secondary	Not relevant to minor	Primary
Walls and steep slopes—reinforcement	Not relevant to minor	Not relevant to minor	Not relevant	Primary

Table 16.2 Minimum geotextile properties[a] required for geotextile serviceability

Required degree of geotextile serviceability	Grab strength (ASTM D-1682/4632) (N/25 mm)	Puncture strength (modified ASTM D-751/3787) (N)	Burst strength (ASTM D-751/3786) (bar)	Trapezoidal tear (ASTM D-1117/4533) (N)
Very high	1200	500	30	340
High	800	340	20	225
Moderate	600	180	15	180
Low	400	135	10	135

[a]All values represent minimum average roll values (i.e. any roll in a lot should meet or exceed the minimum values given). Note: these values are normally 20% lower than manufacturers reported typical values.

Minimum values for these, related to serviceability, are set out in *Table 16.2* with additional data shown in *Figures 16.1, 16.2, 16.3* and *16.4*. Properties of geotextiles also need to be considered in relation to sub-grade conditions, cover materials and construction equipment. Information on these topics is provided in *Tables 16.3* and *16.4*.

16.14.3 Separation

A definition of a separator might be a geotextile placed between two different soil types to prevent intermixing. An example of this might be a granular fill placed over a cohesive soil. In this application serviceability should be the main consideration.

16.14.4 Filtration

The main criteria to be considered are:

- pore size of geotextile
- permeability of geotextile (sufficient to transmit water flow from base soil).

Some guidance on these factors is given in *Figures 16.5, 16.6, 16.7* and *16.8*.

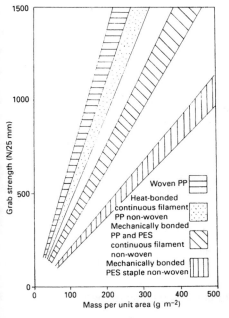

Figure 16.1 *Grab strength (ASTM-D1682) versus mass per unit area. PP, polypropylene; PES, polyester*

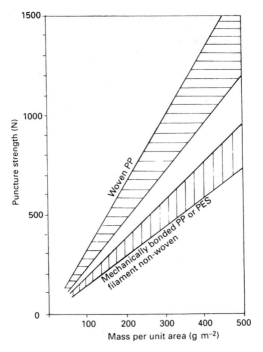

Figure 16.2 *Puncture strength (ASTM D3787) versus mass per unit area. PP, polypropylene; PES, polyester*

16.14.5 Drainage
This may be defined as the ability to transmit fluid in the plane of the geotextile. Of particular use in this application are the thicker non-woven fabrics. Data on the transmissivity of non-woven mechanically bonded geotextiles is shown in *Figure 16.9*.

16.14.6 Reinforcement
The most widely used engineering application of geotextiles is in soil reinforcement. As with other properties, tensile strength and axial stiffness are radically affected by fabric structure and polymer type. This is reflected in *Figure 16.10*.

16.14.7 Economics
The wide variation in cost of various geotextiles is illustrated in *Table 16.5*.

16.15 Natural and synthetic rubbers

Rubbers or elastomers are a large and growing class of materials. At one end of the spectrum, natural rubber and synthetics

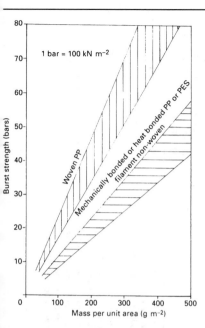

Figure 16.3 *Burst strength (ASTM D3786) versus mass per unit area, PP, polypropylene; PES, polyester*

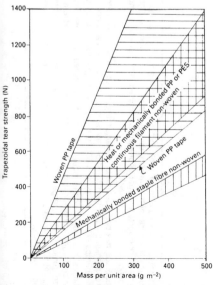

Figure 16.4 *Trapezoidal tear strength (ASTM D1117) versus mass per unit area. PP, polypropylene; PES, polyester*

Table 16.3 *Required degree of geotextile serviceability as a function of subgrade conditions and construction equipment*

| Subgrade conditions | Construction equipment and 150–300 mm initial cover to geotextile | | |
	Low ground pressure equipment ($<30\,kN\,m^{-2}$)	Medium ground pressure equipment ($>30\,kN\,m^{-2}$ $<60\,kN\,m^{-2}$)	High ground pressure equipment ($>60\,kN\,m^{-2}$)
Subgrade has been cleared of all obstacles except grass, weeds, leaves and fine wood debris. Surface is smooth and level such that any shallow depressions and humps do not exceed 150 mm in depth or height. All larger depressions are filled. Alternatively, a smooth working table may be placed	Low	Moderate	High
Subgrade has been cleared of obstacles larger than small to moderate sized tree limbs and rocks. Tree trunks and stumps should be removed or covered with a partial working table. Depressions and humps should not exceed 500 m in depth or height. Larger depressions should be filled	Moderate	High	Very high
Minimal site preparation is required. Tress may be felled, delimbed and left in place. Stumps should be cut to project not more than 150 mm above subgrade. Geotextile may be draped directly over the tree trunks, stumps, large depressions and hump, holes, stream channels and large boulders. Items should be removed only if placing the geotextile and cover material over them will distort the finished road surface	High	Very high	Not recommended

Table 16.4 Required degree of geotextile serviceability as a function of cover material and construction equipment

	Initial lift thickness (mm)					
	150–300		300–450		450–600	>600
	Low ground pressure equipment (<30 kN m⁻²)	Medium ground pressure equipment (>30 <60 kN m⁻²)	Medium ground pressure equipment (>30 <60 kN m⁻²)	High ground pressure equipment (>60 kN m⁻²)	High ground pressure equipment (>60 kN m⁻²)	High ground pressure equipment (>60 kN m⁻²)
Fine sand to 50 mm maximum size gravel, rounded to subangular	Low	Moderate	Low	Moderate	Low	Low
Coarse aggregate with diameter up to one-half proposed lift thickness, may be angular	Moderate	High	Moderate	High	Moderate	Low
Some to most of aggregate with diameter greater than one-half proposed lift thickness, angular and sharp edges, few fines	High	Very high	High	Very high	High	Moderate

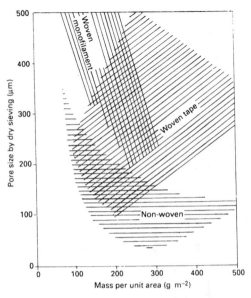

Figure 16.5 *Effect of geotextile structure on pore size versus mass per unit area*

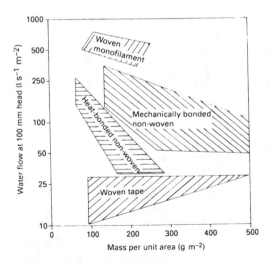

Figure 16.6 *Effect of geotextile structure on water flow versus mass per unit area*

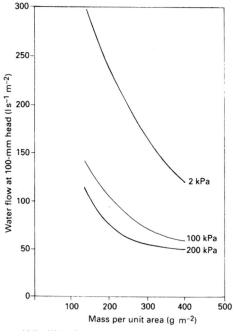

Figure 16.7 *Water flow versus mass per unit area. Data for normal stress levels in the range 2–2000 kPa for mechanically bonded continuous-filament polypropylene non-woven*

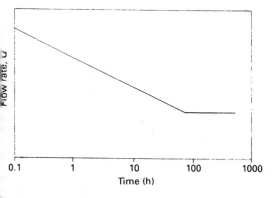

Figure 16.8 *Variation of flow rate with time for a soil-geotextile system*

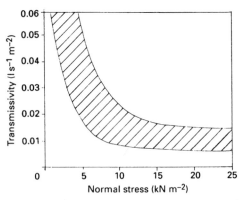

Figure 16.9 *Effect of normal stress level on the transmissivity of non woven mechanically bonded geotextiles.*

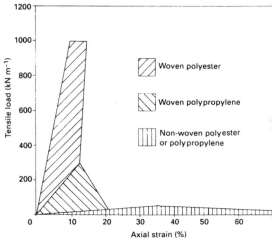

Figure 16.10 *Generalized load strain characteristics (regardless of mass per unit area); areas of overlap have been omitted for clarity*

such as polychloroprene are soft materials which have the ability to undergo extensions of many hundreds of percent yet return rapidly to their original shape when the deforming force is removed. The compliance of rubbers in shear and extension coupled with their near incompressibity means that particular techniques are often required in the stress analysis of these materials. Natural rubber (NR) was the first elastomer and is still the most heavily used. Its mechanical properties epitomize those of a fully rubbery material. Ethylene propylene rubber (EPM/EPDM) is a synthetic elastomer extensively used

Table 16.5 Relative cost per unit tensile strength

Geotextile structure and polymer type	Mass per unit area per unit strength $(g\,m^{-2})/(kN\,m^{-1})$	Strength per unit mass $(kN\,m^{-2})kg^{-1}$	Relative cost per unit strength
Non-woven mechanically bonded polyester continuous filament	12.5	80	1.0
Non-woven mechanically bonded polypropylene continuous filament	15.4	65	1.0
Non-woven heat-bonded polypropylene/polyethylene continuous filament	15.4	65	1.0
Non-woven heat-bonded polypropylene continuous filament	14.3	70	0.9
Woven polyester multifilament	2.3	440	0.6
Woven polypropylene tape yarn	3.8	260	0.4
Woven polypropylene tape	5.9	170	0.4

in construction because of its excellent resistance to weathering. The strength, and related properties, of NR and of EPDM are compared with those of other materials in *Table 16.6*.

In some applications full rubbery behaviour offers great advantages, dock fenders being one example. In other cases, such as window-frame seals and roofing membranes, only a limited degree of elasticity is required. In the latter category there is considerable overlap between elastomers, which are sometimes modified to make them more like thermoplastics, and thermoplastics which have been modified to make them more rubbery.

At the opposite end of the spectrum from natural rubber lie materials which are closely related chemically to rubbers but which are not physically rubbery. Examples include the ebonites, which are hard materials obtained by the vulcanization (heating) of natural or certain other rubbers containing very high levels of sulphur. The latter class of materials lies outside the scope of this chapter.

Additives make an important contribution in the performance of finished elastomers. Without the reinforcing action of fine particles of carbon (carbon black) or silica, some synthetic rubbers would be so weak as to be of little use. Without the protective action of added antioxidants and antiozonants, the surface of natural rubber and other rubbers based on unsaturated hydrocarbon chains would deteriorate within 1 year or so in many environments. Nevertheless, the base polymer of an elastomer remains of key importance.

Useful guidance on symbols and abbreviations may be obtained from ISO 1629. *Table 16.7* indicates a range of common elastomers; *Table 16.8* a list of natural rubber formulations.

16.15.1 Types of rubbers and their properties

The following sections list some but not all of the available rubbers and a selection of their properties. The choice of a rubber for a particular application will usually require expert assistance.

Natural rubber (NR) NR's combination of inherent strength and high elasticity mean that, mechanically, it sets the standard for other elastomers. Typical operating temperature range −60 to +80°C.

Synthetic *cis*-polyisoprene (IR) IRs are, chemically, the closest of the synthetic rubbers to NR. Typical operating temperature range −60 to +80°C.

Butyl (IIR) The damping of IIRs is typically much higher than that of NR and it is this attribute which is exploited in rubber mountings and similar applications. Typical operating temperature range −60 to +120°C.

Styrene–butadiene (SBR) Like NR, SBRs are vulnerable to mineral oils and often require protection against oxidation and ozone attack. Typical operating temperature range −40 to +80°C.

Butadiene (BR) BRs are commonly available with low, medium and very high (>96%) contents of *cis*-butadiene monomeric units. They frequently require protection against oxygen and are vulnerable to mineral oils. Mechanical damping can be lower than that for NR. Typical operating temperature range −70 to +80°C.

Nitrile (NBR) NBRs were some of the first oil-resistant elastomers. They have better resistance than does chloroprene to oils and can be more resistant to heat ageing, although they can be attacked by ozone. Typical operating temperature range −20 to +110°C.

Ethylene–propylene (EPM) EPMs are immune to ozone cracking and attack by diatomic oxygen. They have good tolerance to water, but not to mineral oils. Typical operating temperature range −40 to +130°C.

Chloropene (CR) CRs have greater tolerance than NR to high temperature, mineral oils and oxygen. They may have reasonable fire retardancy but be highly toxic. Typical operating temperature range −30 to +100°C.

Polyurethane (AU or EU) This represents a particularly wide range of materials and thus it is difficult to make general performance statements. Many are strong materials, very much stiffer, less elastic and possessing higher mechanical damping than NR. Typical operating temperature range −30 to +80°C.

Silicone These are noted for their wide operational temperature range and excellent resistance to oxygen and good fire resistance. Typical operating temperature range −90 to +230°C.

Chlorosulphonated polyethylene (CSM) CSMs are notable for their ability to produce brilliant white, durable vulcanizates which are resistant to water and oxygen. Typical operating temperature range −20 to +120°C.

Chlorinated polyethylene (CM) CMs are resistant to oxygen and water; they have a fair tolerance to mineral oils. Typical operating temperature range −20 to 100°C.

Ethylene–methylacrylate (AEM) AEMs give high mechanical damping over a wide temperature range; they have generally good resistance to oxygen and water and are moderately

Table 16.6 Comparison of material properties (ranges are given where appropriate)

Property	Units	Natural rubber				EPDM filled[e]	Polyethylene (high density)	Nylon 66	Aluminium	Mild steel	Glass (common)	Concrete	Wood[f] (oak)	Water
		Unfilled[a]	Filled[b]	Ebonite[c]	Foam[d]									
Density	$Mg\,m^{-3}$	0.95	1.12[i]	1.08–1.10	0.1–0.6	1.2	0.94–0.97	1.14	2.70	7.75	2.4–2.7	2.1	0.74–0.77	1.00
Thermal properties														
Specific heat	$J\,g^{-1}\,°C^{-1}$	1.83	1.50	1.38	1.89	2.09[g]	2.3	1.7	0.95	0.48	0.84	1.89	1.37	4.19
Thermal conductivity	$W\,m^{-1}\,°C^{-1}$	0.15	0.28	0.16	0.045	0.39	0.45–0.52	0.25	230	46	1.1	1.2	0.38	0.65
Coefficient of linear expansion ($\times10^5$)	$°C$	22	18	6	5–10	1.89	11–13	8	2.3	1.1	0.9–1.1	1.0–1.5	0.5[l]	7
Mechanical properties														
Hardness														
Brinell	$kgf\,mm^{-2}$	≪1	≪1	11–12	–	–	1	–	22	140	–340	–	–	0
International rubber scale	IRHD	45	65	100	<20	65	98	100	100	100	100	100	100	0
Tensile strength	MPa	24[k]	24[k]	40–60	0.2–1.5	17	20–35	60–80	70–100	400–500	30–90	20–50	40 120	0
Elongation at break	%	700	550	3–8	200–300	450	20–600	60–100	5	40	3	–	–	0
Young's modulus, E	MPa	1.9	5.9	3000	0.05–0.08	5.9[g]	–1000	1000–3000	70000	2.1×10^5	60000	25000–35000	9000	0
Acoustic properties														
Bulk modulus, K	MPa	2000	2000	4170	–	–	3300	–	70000	1.8×10^5	37000	–	–	2050
Poisson's ratio, v	–	0.499	0.499	0.2	0.33	0.499	0.38	0.38	0.345	0.291	0.22	0.2	–	0.5
Velocity of sound[m]	ms^{-1}	1600	1700[g]	–	–	–	–	–	6400	6000	5000–6000	4300–5300	4000[g]	1500
Electrical properties														
Volume resistivity	Ωm	10^{14}	10^{10}[n]	10^{14}	–	10^{10}[g,h]	10^{15}	10^{12}	10^{-8}	10^{-8}	10^9–10^{11}	10^7–10^{10}	10^9–10^{11}[b]	10^2–10^5[b]
Dielectric constant (relative permittivity)	–	2.5–3.0	15	2.8	–	–	2.25–2.35	4.0–4.6	–	–	7	–	3–6	~80
Dissipation factor	–	0.002–0.04	0.006–0.14	0.004–0.009	–	–	<0.005	0.01–0.02	–	–	0.005	–	0.04–0.05	–

a Gum vulcanizate containing vulcanizing ingredients only.
b Vulcanizate containing c. 50 parts of a reinforcing carbon black per 100 parts, by weight, of natural rubber.
c Unfilled ebonite containing c. 40 parts of sulphur per 100 parts of natural rubber. Ebonite, which is also known as 'hard rubber' in the USA, has a maximum glass transition temperature of c. 80°C, whereas unmodified natural rubber has a transition temperature as low as −70°C.
d Properties of latex foam and other cellular products from natural rubber will vary with extent of foaming or expansion. Young's modulus of open-cell foam can be estimated from Young's modulus of the solid rubber component and the density of the foam.
e Typical formulation (120 parts of black, 95 parts of oil) for roof sheeting.
f Wood is strongly anisotropic. Where appropriate, properties refer to tests along grain.
g Estimated value.
h Across grain, c. 0.2 W m^{-1} °C^{-1}.
j Across grain, c. 5 × 10^{-5}
k Hardness scale range is 0 (e.g. liquid) to 100 (e.g. glass).
l Calculated on the original unstrained cross-sectional area; when calculated using on the cross-section just before break, the value can be as high as 200 MPa.
m For isotropic materials E is related to G and K as follows: $E = 2G(1 + v)$ and $E = 3K(1 - 2v)$.
n Compression varies in medium of large extent.
n Depends on filler type and concentration. High values are obtained by replacing carbon black with silica. Using specially conductive carbon black values under 10Ω m can be achieved.
o The higher value applies to paraffinic timber.
p For distilled water.

Table 16.7 *Available forms of common elastomers*

Elastomer	Bale/ slab	Chip/ pellet	Particulate	Latex	Liquid/ paste	Other
Natural (NR)	**a		*	*	*	
Synthetic cis-polyisoprene (IR)	**a			*	*	
Butyl (IIR)	**			*	*	
Styrene-butadiene (SBR)	**b		*			
Butadiene (BR)	**a			*	*	
Nitrile (NBR)	**	*	*	*	*	
Ethylene-propylene (EPM/EPDM)	**a	*	*			*c
Chloroprene (CR)		**	*	*		
Polyurethane (AU/EU)	*				**	
Silicone (MQ, etc.)	*d				*	
Chlorosulphonated polyethylene (CSM)		*				
Ethylene acrylic (AEM)	*d					
Polyacrylic (ACM/ANM)	*	*	*	**		
Fluorinated hydrocarbon (FPM or FKM)		**			*	
Polysulphide (T)	**				*	
Epichlorohydrin (CO/ ECO)	*					
Polypropylene oxide (GPO)	*					
Polynorbornene	*a		*			*d

*Available form; ** Very common form.
aOil-extended grades are available.
bOil-extended SBR is common. Carbon black filled master batches with and without oil are also available.
cCrumb and friable bale forms also available.
dOften partly compounded.

tolerant of mineral oils. Typical operating temperature range −40 to +150°C.

Fluorinated hydrocarbon (FPM or FKM) FPM/FKMs have excellent resistance to mineral oils, oxygen and high temperatures; they are reasonably tolerant of water. The density of FKM is about 50% higher than that of NR. Typical operating temperature range −20 to +230°C.

Polysulphide Polysulphides are noted for their resistance to a wide range of fluids including mineral oils and water; they also have a high oxygen tolerance but low gas permeability. Typical operating temperature range −40 to +80°C.

Epichlorohydrin (ECO) ECOs have good resistance to oxygen and many mineral oils and can give low damping. They show some vulnerability to water. Typical operating temperature range −10/20 to +130°C.

Table 16.8 *Natural rubber formulations*

Formulation	Parts by weight	
EDS24		
Natural rubber, SMR CV	100	
Zinc oxide	5	
Stearic acid	2	
Carbon black, GPF (ASTM N-660)	20	
Process oil	2	
Antioxidant/antiozonant, HPPD	3	
Antiozonant wax	2	
CBS (accelerator)	0.6	} Conventionally
Sulphur	2.5	} accelerated normal sulphur level, vulcanizing system
EDS4		
Natural rubber, SMRL	100	
Zinc oxide	5	
Zinc 2-ethylhexanoate	1	
Carbon black, SRF (ASTM N-762)	30	
Antioxidant, TMQ	2	
Antiozonant, DOPD	4	
MBS (accelerator)	1.7	} Semi-EV,
TBTD (accelerator)	0.7	} moderately
Sulphur	0.7	} low sulphur, system
EDS42		
Natural rubber, SMR CV	100	
Zinc oxide	5	
Stearic acid	2	
Carbon black, FEF (ASTM N-550)	20	
Process oil	2	
Antioxidant/Antiozonant, HPPD	3	
Antiozonant wax	2	
MBS (accelerator)	2.1	} EV, low
Sulphur	0.25	} sulphur
TMTD (accelerator)	1.0	} system

Polypropylene oxide (GPO) GPO is resistant to oxygen attack; fairly tolerant of water and mineral oils. Can give low mechanical damping. Typical operating temperature range -50 to $+130°C$.

Ethylene–vinyl acetate (EVA) EVAs have good resistance to oxygen and water; fair resistance to petroleum-based oils. Typical operating temperature range -40 to $+120°C$.

Polynorbornene (PNR) PNRs may be very soft materials but with reasonable strengths. They are generally tolerant of water but vulnerable to many mineral oils and oxygen. Typical operating temperature range -40 to $+80°C$.

Thermoplastic elastomers (TPE) In general, thermoplastic elastomers (TPEs) show less complete immediate recovery after deformation than do many conventional elastomers and their creep rates also tend to be higher. Harder materials can be obtained with TPEs than is normal for most conventional elastomers (even polyurethanes); however, for these high modulus TPEs yield strains may be only *c.* 10%. Among the TPEs block co-polymer types (e.g. polyether–polyamide, polyether–polyester and polyurethane TPEs) are available which have Young's moduli of many hundreds of MPa.

16.16 Acrylics and polycarbonates

16.16.1 Acrylics

Acrylics are normally thermoplastic substances, soluble in organic solvents. They vary from polymethylmethacrylate (Perspex), a hard, glass-clear material, to soft rubbery polymers or waxes. Certain water dispersible acrylic polymers are used to modify the workability of specialist concrete repair materials. Methacrylate polymers were first made in 1877, with polymethylmethacrylate first made in 1880. Commercial production of polyacrylates began in about 1927 by Rohm and Haas.

16.16.1.1 Chemical properties   Acrylics tend to be soluble in most aromatic and chlorinated hydrocarbons, esters, ketones and tetrahydrofuran. When cross-linked, they can be insoluble, but will swell in chlorinated hydrocarbons.

They are plasticized by some ester type plasticizers (e.g. dibutylphthalate and tritolylphosphate), are swollen by alcohols, phenols, ether and carbon tetrachloride, but are relatively unaffected by aliphatic hydrocarbons, concentrated alkalis, most dilute acids and concentrated hydrochloric acid, aqueous solutions of salts and by oxidizing agents. They are, however, decomposed by concentrated oxidizing acids and by alkalis in alcoholic solution.

16.16.1.2 Thermal properties
Specific heat: $1.47 \, \text{J g}^{-1} \, \text{K}^{-1}$
Conductivity: *c.* $0.19 \, \text{W m}^{-1} \, {}^{\circ}\text{C}^{-1}$
Coefficient of linear expansion: $0.7 \times 10^{-4} \, {}^{\circ}\text{C}^{-1}$ (20°C); $1.05 \times 10^{-4} \, {}^{\circ}\text{C}^{-1}$ (80°C).
Maximum service temperature: *c.* 80°C.
Decomposes at 180–190°C.

16.16.1.3 Mechanical properties
Young's modulus: *c.* 2940 MPa (decreasing to 1670 MPa at 80°C).
Tensile strength: *c.* 70 MPa (polymethylmethacrylate) falling with increasing temperature.
Elongation at break: *c.* 4% (cast sheet).

Impact strength:
  Notched Izod: $1.6\,kJ\,m^{-2}$.
  Charpy unnotched: $15\,kJ\,m^{-2}$
Compressive strength: 117 MPa.

For elastic modulus, allowable service stresses and coefficients of thermal expansion see *Tables 16.9, 16.10* and *16.11*.

16.16.2 Polycarbonates
Polycarbonates are transparent, faintly amber coloured, thermoplastic materials showing good dimensional stability, thermal resistance and electrical properties, and also good tensile and impact strength. They are used in mouldings fibres and films.

Diphenylolalkanes, the precursors of polycarbonates were first made by A. Bayer in 1872. A. Einhorn made the first polymers in 1898. In the 1950s, the polymers were developed commercially by Farbenfabriken Bayer AG, the General Electric Company in the USA and also by others.

16.16.2.1 Chemical properties Polycarbonates are dissolved by some chlorinated hydrocarbons (e.g. chloroform and methylene chloride). They are swollen by acetone, benzene and carbon tetrachloride. Aromatic hydrocarbons, ketones and esters may cause crazing and stress cracking. Polycarbonates are resistant to aliphatic hydrocarbons, trichloroethylene, alcohols (although prolonged exposure may cause surface crazing), dilute acids and oxidizing agents. They are decomposed by hot alcoholic alkalis, organic amines and are attacked by aqueous alkalis (which may be found in some cleaning fluids).

Table 16.9 *Modulus of elasticity (in MPa) vs. time and temperature*

Temperature (°C)	Time of application of load (days)			
	1	10	100	1000
*Acrylic**				
25	2480	2140	1875	1635
40	2240	1860	1530	1255
50	2170	1725	1380	1105
60	2000	1515	1165	880
70	1860	1380	1000	740
80	1655	1140	760	–
90	1365	760	–	–
Polycarbonate†				
25	2400	2290	2185	2090
40	2055	1875	1710	1560
50	1885	1715	1565	1430

*These values can be used to calculate long-term deflection under continuously applied loads. To calculate momentary or initial deflection under long-term loads, at room temperature, use a value of 3100 MPa.
†Modulus values below 700 MPa have shown wide variation.

Table 16.10 *Maximum allowable service stress (MPa)*

Temperature (°C)	Loading	
	Continuous	Intermittent
Acrylic		
25	10	21
33	9	14
40	8	10
50	7.5	7.5
60	7	7
68	5.5	5.5
80	4	4
90	2	2
25 (FEL)	35	35
Polycarbonate		
−54	29	36
0	16	29
23	14	28
54	11	24
71	7	22
93	3.5	21
121	0	17
23 (FEL)	7	7

FEL, fatigue endurance limit.

Table 16.11 *Some coefficients of thermal expansion*

Material	Coefficient (mm mm^{-1} °C^{-1})
Cast acrylic sheet	0.000074
Polycarbonate sheet	0.000068
Aluminium	0.000023
Steel	0.000011
Plate glass	0.000009

16.16.2.2 Thermal properties

Specific heat: 1.3 J g^{-1} K^{-1}.

Conductivity: 0.20 W m^{-1} °C^{-1}.

Coefficient of linear expansion: 0.6–0.7×10^{-4} °C^{-1} (slightly higher above 60°C).

Maximum service temperature: up to 135°C, although at the higher temperature the polymers may turn brown by oxidation. (NB: there is some creep at 100°C.)

Decomposes at 310–340°C (softens above 160°C, melts at 215–230°C).

16.16.2.3 Mechanical properties

Young's modulus: *c.* 2450 MPa (significantly improved by incorporation of glass fibres).

Tensile strength: up to 70 MPa, yield at about 60 MPa (up to 140 MPa with the incorporation of glass fibres).
Elongation at break: 60–100% (yielding at 5% strain).
Impact strength:
  Notched Izod: 10 kJ m^{-2}.

16.17 Polymer dispersions

16.17.1 Introduction
Polymers are available in solid, powder and liquid forms. The liquids may be solutions, dispersions or 100% polymer. In the context of this section a polymer dispersion is a stable suspension of solid, submicroscopic polymer particles in an aqueous medium, which on drying out under the conditions of use forms a coherent film. Dispersions are often erroneously called emulsions, a term which should be reserved for two-phase suspensions where both phases are liquid. Polymer dispersions are also referred to as latices (or latexes) hence the terms latex modified concrete and latex modified mortar.

Polymer dispersions are milky white liquids which come in a wide range of viscosities and solid content. They are typically supplied at 40–70% solids, but may be supplied in a more dilute form, e.g. by diluting with water for a particular end-use such as prepack concrete-repair systems. The most common polymer dispersion product in everyday life is emulsion paint, most varieties of which should correctly be called aqueous dispersion paint.

Typical uses of polymer dispersions in cementitious mixes include:

(1) Bonding agents
(2) Floor screeds, underlayments and toppings
(3) Concrete repair mortars
(4) Fixing brick slips and ceramic tiles
(5) Bedding paviors
(6) Bricklaying mortars and external renders
(7) Corrosion protection of steel
(8) Renders over external insulation
(9) Road-bridge decks
(10) Sprayed coatings
(11) Pipe coatings
(12) Grouting of open-texture macadam.

Table 16.12 gives an indication of the type of dispersion that may be used in particular applications.

16.7.2 Properties
It is beyond the scope of this book to provide a wide range of properties for dispersions since they themselves vary widely in both formulation and application. However, an indication of the variation in properties of cement mortar can be gained from *Table 16.13*.

Table 16.12 *Some examples of the selection of dispersion type according to application*

Application	Conditions	Dispersion type[a]
Adhesives[b]	Dry	PVA, EVA
	Wet or dry	SBR, SA, CR
Bitumen modification		NR, SBR, CR
Bonding agents[b]	With cement, wet or dry	SBR
	Without cement, dry	PVA, EVA, PAE
	Without cement, wet or dry	SBR, PVDC
Primer paints for steel		SBR, PVC, PVDC[c]
External coatings		PAE, SA
Flexible coatings	With cement; very high flexibility	NR, PAE
Floors for animal husbandry; rubber-crumb mixes		NR, NR SBR
Mastics		PAE
Mortar/concrete/floor screeds and toppings	Dry	PAE, SA
	Wet or dry	SBR
Oil-resistant products		CR, NBR
Ship-deck screeds		NR, SBR, PAE
Soil stabilization		SBR

[a]PVA, polyvinyl acetate; EVA, ethylene vinyl acetate; SBR, styrene butadiene; SA, styrene acrylics; CR, chloroprene; NR, natural rubber latex; PAE, polyacrylic ester; PVDC, polyvinylidene dichloride.
[b]In this contect adhesives are used for joining dry solids and do not contain cement. Bonding agents are products which assist the bond of wet mixes to the background.
[c]Often modified with acrylics.

16.8 Polyethylene

Polyethylene is a thermoplastic and is produced by polymerizing ethylene using commercially developed processes. A list of typical production processes is given in *Table 16.14*. The properties of the material are dependent upon three parameters (see *Table 16.15* and *Figure 16.11*):

- melt flow rate (MFR), which depends on the molecular weight
- density, which depends on the crystallinity
- the molecular-weight spread, which depends on the polydispersity (see footnote to *Table 16.15*).

Types and usage of polyethylene are:

(1) LDPE (low density)
(2) MDPE (medium density)
(3) HDPE (high density)
(4) LLDPE (linear low density).

These descriptions can, however, be somewhat misleading and the reader is referred to *Table 16.14* for more accurate

Table 16.13 Typical physical properties of mortars modified with SBR dispersion[a]

	% Polymer dispersion by weight of cement, i.e. wet latex on cement							
	0	10	15	20	25	30	40
Water : cement	0.40	0.34	0.33	0.31	0.28	0.25	0.24
Compressive strength (N mm^{-2})	55			50			
Tensile strength (N mm^{-2})	4	5	6	7	8	8.5	9.0
Flexural strength (N mm^{-2})	10.0	11.5	14	15	16	18	19

[a]SBR was Revinex 29Y40 (47% solids). PAE (e.g. VP4001) and SA (e.g. BASF S702) gave similar results at 20% dispersion on cement, and at similar mix density. The mortar was a floor topping mix consisting of 1 : 3 OPC : coarse washed sand (BS 882: 1983, complying with both the M and C grading) by weight. All mixes of similar dry consistency. Curing was 1 day covered in the mould and 27 days air curing. Test methods in accordance with BS 6319: Parts 2, 3 and 7. The strength results on repair mortars are likely to be up to 50% lower, as a more workable mix is required. However, the proportional increases, versus the unmodified mix, will be maintained.

Table 16.14 *Density and melt flow rate (MFR) ranges for polyethylene process types*

Process type	Density (kg m⁻³)	MFRᵃ (g/10 min)
High pressure, autoclave or tubular reactor	910–928	0.1–7.0
Slurry (suspension), low/medium pressure, Phillips catalysts	930–965	2ᵇ–4.0
Slurry (suspension), low/medium pressure, Zeigler catalysts	930–965	<1ᵇ–50.0
Gas phase, proprietary catalysts	910–960	<0.1–100
Solution (liquid phase), Phillips/Zeigler	900–960	0.25–80.0

ᵃ2.16 kg at 190°C (ISO 1133).
ᵇLoad 21.6 kg.

Table 16.15 *Typical effect of molecular weight (MW), crystallinity and molecular weight distribution* (MWD) on the properties of polyethylene*

Property	Increase in average MW	Increase in crystallinity	Increase in breadth of a given MWD
Melt flow rate	Decrease	–	–
Density	–	Increase	–
Rate of change of bulk melt viscosity with shear rate	–	–	Increase
Tensile strength at yield	–	Increase	–
Tensile strength at break	Increase	Increase	–
Elongation at break	Increase	Decrease	–
Impact strength	Increase	Increase	Decrease
Resistance to deformation	–	Increase	–
Barrier properties	–	Increase	–
Chemical resistance	Increase	Increase	–
Dielectric constant	–	Increase	–
Stiffness	–	Increase	–
Hardness	–	Increase	–
Environmental stress crack resistance	Increase	Decrease	Increase
Thermal conductivity	–	Increase	–

ᵃAlso known as polydispersity.

definitions. The considerable use of this material is indicated in *Table 16.16*.

16.8.1 Properties

The general physical properties are indicated in *Table 16.17*. However, considerable property variations can be achieved by varying the formulation of the material. Particular properties for polyethylene used for yellow gas pipe work are shown in

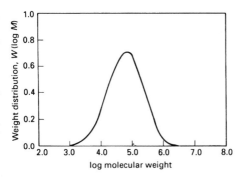

Figure 16.11 *A typical normal molecular-weight distribution of polyethylene. Averages: $M(N) = 25.7$; $M(V) = 105.1$; $M(V) = 128.5$; $M(Z) = 448.9$; polydispersity $= M(W)/M(N) = 4.99$*

Table 16.16 *The usage of polyethylene in the USA, Western Europe and Japan in 1988*

Application	Million tonne
Blow moulding	2.8
Film/coating	8.6
Injection moulding	2.8
Pipe/cable extrusion	1.3

Table 16.18; general properties for hot-water pipe are shown in *Table 16.19*. Further properties for polyethylene used as a foamable insulation compound and as a solid skin compound are shown in *Tables 16.20* and *16.21* respectively.

For damp-proof membrane applications useful information can be found in *Low density polyethylene film for building applications, Parts 1 and 2*, published by the Packaging and Industrial Films Association.

16.19 Polystyrene

Synthetic polystyrene was first made over 125 years ago. However, it was not until 1936 that the first commercial production became available. In that year some 800 tonne was produced in Germany increasing to 5000 tonne by 1942.

In its solid state, polystyrene is a clear non-crystalline plastic material which is brittle. Improvements to its impact strength can be made by compounding with more rubbery polymers. These high-impact grades can be extruded or moulded into products for use inside buildings; for example in light fittings when adequate ultraviolet light resistance is needed to cover

Table 16.17 *Physical properties of polyethylene (general)*

Property	Method	Units	Polyethylene		
			Low density	Medium density	High density
Density	ISO 1183	kg m^{-3}	910–925	926–940	941–965
Tensile strength at yield	ISO 527 (type 2 speed)	MPa	8–13	13–19	19–32
Elongation at break	ISO 527 (type 2 speed)	%	50–200	50–1000, depending on molecular weight (MFR)	700–1300
Flexural modulus	ISO 178	MPa	50–200	200–700	700–1300
Impact strength (Charpy)	ISO 179	kJ m^{-2}		6 to non-break	
Vicat softening point, 1 kg load	ISO 306	°C	85–100	100–118	118–133
Environmental stress, crack resistance, F_{50} time	ISO 4599	h	<1 to >1000 h, depending on molecular weight (MFR)		
Coefficient of thermal expansion		10^{-5} °C^{-1}		10–13	

Table 16.18 *Physical properties of yellow gas pipe/fittings compound (medium density polyethylene)*[a]

Property	Typical value	Units	Test methods
Melt flow rate 190°C, 2.16 kg	0.2	g/10 min	ISO 1133, Proc. 4
Density	940	kg m^{-3}	ISO 1872
Tensile stress at yield	18	MPa	ISO R527
Elongation at break	>600	%	ISO R527
Impact strength (Izod)	930	J m^{-1} notch	ISO 180
Flexural modulus	700	MPa	ISO 178
Vicat softening temperature	116	°C	ISO 306, method A
Brittleness temperature	< −70	°C	ISO 974
Environmental stress crack resistance, F_{50}	>1000	h	ISO 4599

[a]Courtesy of BP Chemicals Ltd.

Table 16.19 *General hot-water pipe properties*

Property	Units	Value	Method
Density	kg m^{-3}	944	NFT 54-002
Thermal conductivity	W K^{-1}	0.35	DIN 56612
Coefficient of expansion	mm m^{-1} K^{-1}	0.19	DIN 53752
Elongation at breaking	% (100 mm min^{-1})	375	ISO R527
Reversion	%	1.3	ISO 2506
Circumferential stress at 50 years	MPa (20°C)	9.87	ATEC of CSTB
	MPa (40°C)	7.05	ATEC of CSTB
	MPa (60°C)	6.45	ATEC of CSTB
	MPa (80°C)	3.90	ATEC of CSTB

fluorescent tubes or as a lightweight substitute for glass in shower cubicles or screens.

It is, however, as a foamed plastic that polystyrene is mainly used in construction. Techniques for producing suitable foams were not available until 1952. By the mid-1960s the benefits of the material were more widely recognized and applications began to be developed. Today, polystyrene foams are the most widely used of the plastic insulants available.

Two basic processes are available to produce foams from polystyrene. To simplify the nomenclature we will use the following abreviations:

PS   a polystyrene foam produced by either an expansion process or an extrusion process

Table 16.20 *Some general physical properties of a foamable insulation compound*[a]

Property	Units	Typical value	Method
Density	kg m^{-3}	942	Adaption of ISO 1872/1[a]
Tensile strength of cellular insulation	MPa	14	
Resistance to compression	N	175	
Resistance to thermal ageing at 105°C	h	1500	British Telecom Specification M142D
Dielectric loss angle of 1 MHz at 23°C	μR	400	
Dielectric constant of 1 MHz at 230°C		1.74	ASTM D1531: 1975

[a]All samples except for density were cellular insulation of 0.15 mm radial thickness, expanded to a specific gravity of 550–600 kg m^{-3} on a 0.5 mm diameter copper conductor.

Table 16.21 *Solid skin compound properties*

Property	Units	Typical value	Method
Melt flow rate	g/10 min	0.6	ISO 1133, Procedure 4
Density	kg m^{-3}	945	ISO 1872: Part 1
Oxidative induction time at 200°C	min	40	–
Petroleum jelly absorption at 10 days at 70°C	% gain	6	FT$_z$72 TV1
Dielectric constant		2.3	At 1 MHz ASTM D1531: 1975
Dielectric loss angle	μR	200	At 1 MHz

EPS   a polystyrene foam manufactured by moulding blocks boards or shapes from expandable beads, sometimes referred to as a 'bead-board'

XPS   an extruded polystyrene foam manufactured into boards.

16.19.1 Properties

The physical property requirements of PS, EPS and XPS are shown in *Tables 16.22*, *16.23* and *16.24* respectively. Some of these materials for use in particular applications are covered by BBA or WIMLAS Certification (see *Table 16.25*).

Table 16.22 *Physical properties of solid polystyrene*

Density ($kg\,m^{-3}$)	1050
Melting point (K)	510
Specific heat capacity ($J\,kg^{-1}\,K^{-1}$)	1300
Coefficient of thermal expansion ($mm^{-1}\,mK^{-1}$)	0.070
Thermal conductivity ($W\,m^{-1}\,K^{-1}$)	0.08
Tensile strength (MPa)	50
Young's modulus (MPa)	3100

16.20 Polytetrafluoroethylene (PTFE)

Polytetrafluoroethylene (PTFE) is a wax-like solid having a greasy feel. It is odourless, tasteless, non-flammable and white in colour but can be pigmented. PTFE is a linear chain polymer of great molecular strength and is mainly exploited for its chemical inertness, low coefficient of friction and first class electrical properties. PTFE is not oxidized easily, it is resistant to all common solvents and remains stable at extremes of atmospheric temperatures, but it exhibits plastic deformation under load. PTFE has the largest temperature working range of all plastics and the greatest resistance to chemical attack. Within its working temperature range it is a completely inert product, but at very high temperatures, i.e. in excess of 250°C, it can give rise to decomposition products which have unpleasant effects if inhaled. The properties of PTFE are summarized in *Table 16.26*.

PTFE is of particular interest in offshore work where its low coefficient of friction is beneficial in skid beam systems for launching offshore platforms. Guidance on sliding bearing pressures is indicated in *Table 16.27*.

16.21 Vinyl materials

The use of polyvinyl chloride (PVC) in the construction industry has become established around the world, particularly in the past 30 years. The amount of PVC used for particular applications does vary from country to country for a variety of reasons which may include the type of building methods used, the protection of vested interests, the availability of traditional materials and many others. However, if the European market for PVC in building is surveyed, it will be seen that this major thermoplastic has achieved an important status. By far the largest single use of PVC in the construction industry is in the pipe business which, in 1989, consumed 28% of total PVC usage. So it can be seen that the building market is a very important outlet for PVC as well as the fact that PVC provides extremely useful products for this sector. Other important applications include profiles, window frames, sheet products and flooring.

Table 16.23 The physical property requirements of EPS*

Physical property	Grade				
	SD	HD	EHD	UHD	ISD
Maximum thermal conductivity at 10°C (W m⁻¹ K⁻¹)	0.038	0.035	0.033	0.033	0.041
Compressive strength or compressive stress at 10% strain (kPa)					
Minimum	70	110	150	190	25
Maximum	–	–	–	–	45
Minimum cross-breaking strength (kPa)	140	170	205	275	NA
Dimension stability at 80°C					
Maximum percentage linear change	1.0	1.0	1.0	1.0	1.0
Maximum water vapour permeability at 38°C (ng Pa⁻¹ s⁻¹ m⁻¹)	6.9	5.0	4.2	4.2	10.0
Burning characteristics, extent burnt (mm)					
type N			≥ 125		
type A			< 125		
Safe working load at 1% strain (kPa)	21	45	70	100	10

NA, not applicable.
*Methods of test are referred to in BS 3837: Part 1.

Table 16.24 *The physical property requirements of XPS**

Physical property	Skinned					Planed	
	E1	E2	E3	E4	E5	E6	E7
Compressive stress (kPa) (min.)	100	150	200	300	350	100	250
Water vapour permeability (ng Pa^{-1} s^{-1} m^{-1}) (max.)	3.2	2.7	2.4	2.1	1.9	3.8	2.4
Water absorption by immersion (max. % vol.)	1	1	1	1	1	2	2
Water absorption by diffusion (max. % vol.)	–	4	4	3	3	–	–
Thermal conductivity at 10°C (W m^{-1} K^{-1}) (max.)	0.032	0.030	0.030	0.028	0.028	0.035	0.030
Dimensional stability for 2 days at 70°C (change %) (max.)	5	5	5	5	5	5	5
Compressive creep at 40 kPa and 70°C for 7 days (change %) (max.)	–	–	5	5	5	–	5
Burning characteristics of small specimens	–	–	5	5	5	–	5
			Maximum extent of burning 125 mm				

*Methods of test are referred to in BS 3837: Part 2.

Table 16-25 BBA and WIMLAS approvals and their applications

		EPS	XPS
Partial cavity fill		*	*
Drylining laminates		*	*
External wall insulation		*	*
Floor insulation:	Separate components	*	–
	Laminates	*	*
	Structural infill	*	–
Roof insulation:	Warm flat	*	–
	Inverted	–	*
	Pitched sarking board	–	*
Clayheave		*	–

Polyvinyl chloride is the most widely used thermoplastic in the construction industry. The product name is often abbreviated to PVC but this needs to be more carefully qualified to enable the user to understand the nature and properties of the product he is using. Other terms which are often encountered in industry relating to PVC are 'vinyl' and 'UPVC' and, although neither of these terms is now officially sanctioned by ISO nomenclature, they are widely used, especially where the consumer is concerned and, therefore, need to be understood.

The term PVC has now come to mean a material or product made from a PVC composition, i.e. an intimate mixture of a vinyl chloride polymer or copolymer with various additives.

The additives are required to enable processing to take place and include products such as heat stabilizers, lubricants, fillers and pigments, all of which are required to provide the properties required in the final product. There are other additives which are often used, such as processing aids and impact modifiers which also have a process- or property-enhancing effect. The most important group of additives are probably plasticizers, because they can impart to the PVC product a very wide range of flexible properties. Thus there are two types of PVC product used in the building industry: (i) rigid or unplasticized PVC which is based on a composition which does not contain plasticizer; and (ii) flexible or plasticized PVC. These types are defined by ISO. Traditionally, the rigid forms of PVC have been called UPVC but new ISO nomenclature states that such products should be termed PVC-U and flexible PVC compositions and products should be termed PVC-P.

Table 16.28 indicates a comparison between grades of materials. A description of types is shown in *Table 16.29* whilst typical properties of PVC-U and flexible PVC are shown in *Tables 16.30* and *16.31* respectively. The chemical resistance of PVC-U is indicated in *Table 16.32*. Some guidance is given to the performance of vinyl in fire by reference to *Table 16.33*.

Table 16.26 Properties of polytetrafluoroethylene

Property	Value
Physical properties	
Tensile strength	12.0–41.0 N mm^{-2}
Elongation at break	100–600%
Compressive strength	4.0–12.0 N mm^{-2}
Density (at 23°C)	2.13–2.24 g cm^{-3}
Modulus of elasticity	345–620 N mm^{-2}
Flexural strength	No break
Impact strength	0.13–0.21 J mm^{-1} of notch
Hardness (Rockwell)	D50–D65
Water absorption (24 h, 3 mm thickness)	0.01%
Thermal properties	
Coefficient of expansion (varies with temperature)	10×10^{-5} °C
Conductivity	6 cal cm cm^{-2} °C^{-1} s^{-1} $\times 10^{-4}$
Service temperature	−260–+250°C
Transition point	327°C
Specific heat	0.25 cal °C^{-1} g^{-1}
Flammability	Nil
Electrical properties	
Dielectric strength (short-time, 3 mm)	15 750–23 600 V mm^{-1}
Dielectric constant	2.1×10^6 cycle s^{-1}
Power factor	0.0002×10^6 cycle s^{-1}
Volume resistivity (at 23°C and 50% RH^a)	10^{19}
Arc resistance	7300 s
Chemical resistance	
Concentrated inorganic acids	Unaffected
Dilute inorganic acids	Unaffected
Organic acids	Unaffected
Strong alkalis	Unaffected
Weak alkalis	Unaffected
Petroleum products	Unaffected
Solvents	Unaffected
Sunlight	Unaffected
Weather	Excellent
Permeability	
Carbon dioxide	2.60 g m^{-2} day^{-1} mm^{-1}
Water vapour	1.38 g m^{-2} day^{-1} mm^{-1}

^{a}RH, relative humidity.

16.22 Aramids

Aramid fibres are one of the high performance modern fibres that are potentially of interest to civil and structural engineers, being characterized by high strength, high stiffness, low creep and resistance to corrosion. Unlike carbon and glass fibres, however, the aramid fibres are frequently used without resin impregnation, since they are much more resistant to local bending effects.

Table 16.27 *Allowable sliding bearing pressures for pure PTFE*

Design load effects	Maximum average contact pressure ($N\,mm^{-2}$)		Maximum extreme fibre pressure ($N\,mm^{-2}$)	
	Bonded PTFE	Confined PTFE	Bonded PTFE	Confined PTFE
Permanent design load effects	20	30	25	37.5
All design load effects	30	45	37.5	55

The term aramid is used to refer to aromatic polyamides containing chains of aromatic (benzene) rings, linked together with —CO— and —NH— end groups. Many forms can be produced, but those based on *para* links on the aromatic ring generally give the strongest fibres.

These fibres are now available under a variety of commercial trade names, such as Kevlar (manufactured by Du Pont in America and Northern Ireland), Twaron (manufactured by Akzo in The Netherlands) and Technora (manufactured by Teijin in Japan). The fibres have a modulus greater than $40\,kN\,mm^{-2}$, and so fall within the high-performance category. This category effectively excludes conventional textile fibres, nylon and polyethylene, on the basis of either strength, stiffness or creep.

Du Pont produced the aramid fibre Nomex, poly(*m*-phenylene isophthalamide), in 1958, and research increased on other aramid fibres by this company and its competitors. A concentrated research effort by Du Pont led to the discovery of the precursor to Kevlar in 1965. This was known as Fibre B and was based on poly (*p*-benzamide); it may be produced by either wet- or dry-spinning procedures. The development of Fibre B now appears to have been halted, however, probably owing to the high cost of the starting monomer (*p*-aminobenzoic acid) and the limited stability of the spinning dope.

The Kevlar polymer, poly(*p*-phenylene terephthalamide) or PPT, had been prepared as early as 1958, but the existing spinning techniques had failed to produce a high-strength fibre. It was known, however, that PPT polymer dissolved in concentrated sulphuric acid, and so when a technique for spinning from strong acids became available, the preparation of PPT fibres was reconsidered. When the well-known dry-jet wet-spinning process was used in conjunction with a sulphuric acid spinning dope, a PPT fibre was produced with properties that surpassed those of previous developments. Furthermore, this new breakthrough also led to increased productivity and considerable cost savings. Hence, the registered name of Kevlar was announced by Du Pont in 1973 to replace that of Fibre B.

Table 16.28 A comparison of typical grades of PVC-U, polypropylene and high-density polyethylene (HDPE)

Property	Test method/recommendation	PVC-U	Polypropylene	HDPE
Yield stress (MN m^{-2}) at 23°C	BS 2782: 1970 Method 301G	55	25	20
Tensile modulus 100s 1% strain (GN m^{-2}) at 23°C	BS 4618	2.7–3.0	As moulded: 0.8–1.0 Annealed at 140°C: 1.2–1.4	0.7–0.95
Izod impact strength (J m^{-1})*	BS 2782: 1970 Method 306-A	110 (unmodified) 540–800 (impact modified)	110 (homopolymer) 540 (copolymer)	270–1100
Relative density		1.38–1.45	0.90–0.91	0.940–0.965
Maximum continuous service temperature (°C) from field experience		60	110	80
Coefficient of thermal expansion at 20°C (°C^{-1})	BS 4618	6 × 10^{-5}	12 × 10^{-5}	11 × 10^{-5}
Flammability (oxygen index) (%)	ASTM D2863-77 (Fenimore Martin test)	45	17.5	17.5

*J m^{-1} = 1.873 × 10^{-2} ft. lb. in^{-1}.

Table 16.29 Description of types of PVC-U pipe

Type of pipe	Important properties	Relevant British Standards
Pressure pipe (water mains, chemical plants)	High tensile strength, high modulus, toughness, good creep characteristics, good flow characteristics, good chemical resistance	BS 3505 BS 3506 BS 4346 (for fittings)
Soil pipe	High modulus, toughness, good weathering	BS 4514
Rainwater goods	High modulus, toughness, good weathering, non-corrosive	BS 4576
Drain and sewer pipes	High stiffness, good creep characteristics, toughness, high tensile strength	BS 4660 BS 5481
Land-drainage pipes	High stiffness/unit weight, toughness, good weathering	BS 4962
Ducting pipe (e.g. communications ducting)	High stiffness/unit weight, toughness	British Telecom specification
Conduit	High toughness, good insulating properties	BS 6099-2-2 BS 4607

Table 16.30 *Typical properties of high-impact PVC-U window material*

Property	Test method	Typical value
Tensile strength at 23°C	ISO 527 BS 2782: Method 320C	44 MPa
Tensile modulus (1% strain) at 23°C	ISO R899 BS 4618	2250 MPa
Flexural modulus	ISO 178 BS 2782: Method 335A	2400 MPa
Flexural yield strength at 23°C	ISO 178	76 MPa
Relative density at 23°C		1.4–1.5
Charpy impact strength at 23°C	ISO 179	14 kJ m^{-2}
	BS 2782: Part 3: Method 359: 1984 (0.1 mm V-notch) DIN 53453 (U-notch)	40 kJ m^{-2}
Retention of impact strength after accelerated weathering to 8 GJ m^{-2}	BS 2782: Method 359 (0.25 mm double V-notch)	>80%
Colour-fastness	BS 2782: Method 540 B ISO 4892, Xenotest BS 1006: Part A02: 1978 Colour change: 1 = most change, 5 = least change	Grey scale 4–5
Vicat softening point	ISO 306 BS 2782: Method 120 B	81°C
Flammability (oxygen index)	ASTM S2893 (Fenimore Martin)	45%
Fire resistance	BS 476: Part 7: 1971	Class 1 (most resistant)
Coefficient of linear thermal expansion		6 × 10^{-5} °C^{-1}
Coefficient of thermal conductivity at 23°C		0.16 W m^{-1} °C^{-1}
Reversion	1 h at 100°C	1.6%

Table 16.31 *Typical properties of flexible PVC*

Property	Unit	Value
Relative density		1.19–1.68
Tensile strength	MN m^{-2}	7.5–30
Elongation at break	%	140–40
Hardness (Rockwell)		5–100
Brittleness temperature	°C	−20 to −60
Volume resistivity at 23°C	Ω cm	10^{10}–10^{15}
Ageing resistance		Excellent
Ozone resistance		Very good

Table 16.32 *Chemical resistance of PVC-U to some common chemical products*

Chemical	Temperature (°C)	
	20	60
Alcohol (40% aqueous)	++	+
Antifreeze	++	++
Dish-washing liquid	++	++
Detergent (diluted)	++	++
Furniture polishes	++	++
Gas oil	++	++
Lanolin	++	++
Linseed oil	++	++
Mineral oils	++	++
Moist acidic atmosphere	++	+
Moist alkaline atmosphere	++	+
Motor oils	++	++
Petrol	++	++
Sea-water	++	++
Vinegar	++	++
Vegetable oils	++	++

Key: ++, resistant; +, practically resistant.

16.22.1 Properties

Properties acredited to aramids include:

(1) High tensile strength
(2) High stiffness
(3) High specific strength
(4) Low density
(5) Low creep
(6) Finite life when subjected to high stresses
(7) Excellent longitudinal tensile fatigue performance
(8) Good shock-loading performance
(9) Poor compressive strength
(10) Moderate abrasion resistance
(11) Good chemical resistance
(12) Poor ultraviolet resistance
(13) Resistance to high temperature
(14) Relatively good thermal stability
(15) Electrical non-conductivity.

A guide to comparative tensile properties is shown in *Table 16.34* and typical stress–strain curves are compared with that for prestressing steel in *Figure 16.12*.

16.23 Glass reinforced plastics (GRP)

GRP materials are adaptable for use in construction because of their ease of mouldability, high specific strength, lightness in

Table 16.33 *Ignition properties of PVC*

Property	Test method	PVC-U	Flexible PVC	Wood
Flash ignition temperature* (°C)	ASTM D1929	400	330–380	210–270
Self-ignition temperature† (°C)		450	420–430	400
Oxygen index	ISO 4589	50	23–33	21–23
ISO Radiant Core	ISO 5657			
Ignition time at 30 kW m^{-2} (s)		112	50–75	30–90
Ignition time at 50 kW m^{-2} (s)		33	17–26	4–30
Needle flame test	IEC 695-2-2	Non-ignitable	Ignition only at high levels of plasticizer	Ignitable in <20s

*The lowest temperature of air passing around the specimen at which sufficient combustible gas is evolved to be ignited by a small external pilot flame.
†The lowest temperature of air passing around the specimen at which, in the absence of an ignition source, ignition occurs by itself.

Table 16.34 Comparative tensile properties*

	Ultimate tensile strength (N mm^{-2})	Strain at ultimate (%)	Initial modulus (kN mm^{-2})	Specific gravity	Specific strength††
Aramids					
Kevlar 149	2410	1.4	146	1.47	1.64
Kevlar 49	2900	1.9	120	1.44	1.92
Kevlar 29	2900	3.7	58	1.44	1.92
Twaron	2800	3.3	80	1.44	1.94
Twaron HM	2800	2.0	115	1.45	1.93
Technora	3100	4.4	71	1.39	2.23
Non-aramids					
Nylon	990	18.3	6	1.14	0.87
Polyester	1120	14.3	14	1.38	0.81
E-Glass	2500	4.0	70	2.6	0.96
S-Glass	4600	5.5	85	2.5	1.84
Carbon fibre	2200–5400	0.4–1.8	238–444	1.76–1.9	1.13–3.00
PBO	3400	1.0	340	1.5	2.27
Boron fibre	3150	0.8	379	2.49	1.27
Mild steel	300	20.0	200	7.85	0.03
Prestressing steel	1700	1.6	200	7.85	0.22

*Figures in this table should be taken as a guide only. For most fibres, even more than for bulk materials, final properties are heavily dependent on the size of the fibre and the exact details of the production process, as the wide range for carbon fibres makes clear.
†Specific strength = tensile strength/density.

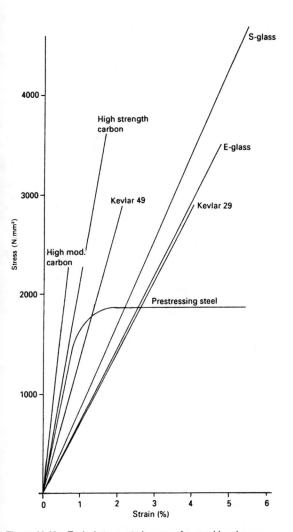

Figure 16.12 *Typical stress–strain curves for aramid and some comparable alternative materials. Note that for most of these materials a wide spread of values can result from changes in details of the production process. Actual fibre tests show slight non-linearities in most cases*

weight, impermeability, non-corrosive nature and the availability of almost infinite variations of colour and finish. In particular, ease of mouldability, high strength and lightness in weight makes the production and handling of physically large products, from flat panels to complex shapes, possible.

Because of these properties these materials have been used in the construction of buildings for nearly 40 years, although the more visible architectural applications have generally grown since the early 1960s.

Over the same time period, GRP materials have also become widely used in civil engineering, particularly in sewage and effluent control products, water storage and distribution, 'tanking' and a great variety of situations where corrosion resistance, impermeability and chemical/environmental resistance is important.

These materials have also become well established over many years as a concrete shuttering/forming medium, particularly where complex shapes and intricate details are required. Also commonly, GRP is used for permanent/*in situ* formwork, often facilitating the use of good decorative external GRP finishes to cement and concrete structures.

More recent materials developments in the area of GRP and 'composites' has led to rapidly increasing interest in these materials for more highly stressed and/or specialized civil engineering applications for which GRP materials may not have been considered entirely suitable in the past.

Examples of GRP applications include:

(1) Architectural forms and features, domes, arches, etc.
(2) External cladding panels and systems
(3) Geodesic structures
(4) Translucent profiled and flat sheeting
(5) Fascias, feature mullions, transoms, etc.
(6) Gutters, downpipes and rainwater systems
(7) Dormer windows
(8) Rooflights, roof panels, mansards, etc.
(9) Feature window and door surrounds
(10) Balconies/balustrades
(11) Canopies, columns, supports, etc.
(12) Decorative and simulated features of all types including coving, cornices, etc.
(13) Plant rooms and enclosures
(14) Modular building and kiosk systems
(15) Walkways
(16) Water storage tanks and cisterns
(17) Septic tanks, sewage services, digesters, etc.
(18) Underground pipe, conduit and duct systems
(19) Concrete formwork of all types
(20) 'Site applied' GRP linings/membranes, etc.

A comparison of the mechanical properties of GRP laminates and those of other structural materials is given in *Table 16.35*.

Table 16.35 *The mechanical properties of GRP laminates compared with those of other structural materials**

Material	Grade	Specific gravity	Elastic modulus (GPa)	Proof strength		Impact strength (kJ/m⁻²)	Specific strength (MPa)	Specific modulus (GPa)
				Tensile (MPa)	Compressive (MPa)			
Mild steel	BS 15	7.8	207	240	240	50	31	27
Aluminium alloy	HE15WP	2.7	69	417	417	25	154	26
Stainless steel	316	7.92	193	241	241	1356	30	24
CSM/polyester	33†	1.47	8	80	120	75	54	5
Unidirectional GRP	82†	2.16	53	900	450	250	417	25

*Courtesy of Scott Bader Co. Ltd.
†Glass content by weight.
CSM, chopped strand mat.

GRP materials as normally used in the construction industry consist of 'syrupy' resins which are applied in layers to a moulding surface, together with alternating layers of glass fibre reinforcement which is usually in the form of a 'mat' of chopped strands.

Typical physical properties for GRP with different types of glass fibre reinforcement appear in *Table 16.36*. The glass content has a significant bearing on the performance of GRP and this is illustrated in *Figures 16.13*, *16.14* and *Table 16.37*. Comparative thermal performance is shown in *Table 16.38*. Care should be taken in using GRP in sandwich laminates; where such techniques are required relative rigidities are indicated in *Table 16.39*.

Table 16.36 *Typical physical properties of GRP with different types of glass fibre reinforcement**

Property	Unit	Chopped strand mat	Woven rovings	Satin weave cloth
Glass content	% wt	30	45	55
	% vol.	18	29	38
Specific gravity		1.4	1.6	1.7
Tensile strength	MPa	100	250	300
Tensile modulus	GPa	8	15	15
Compressive strength	MPa	150	150	250
Bend strength	MPa	150	250	400
Modulus in bend	GPa	7	15	15
Impact strength : izod unnotched†	kJ m^{-2}	75	125	150
Coefficient of linear expansion	$\times 10^{-5}$ °C^{-1}	30	15	12
Thermal conductivity	W m^{-1} K^{-1}	0.20	0.24	0.28

*Courtesy of Scott Bader Co. Ltd.
†Tested edgewide.

Table 16.37 *Effect of glass content on the properties of contact moulded CSM polyester laminates (typical values)**

Resin : glass ratio by weight		2:1	2.5:1	3:1
Glass content (% wt)		33	29	25
Glass content (% vol.)		20	17	14
Specific gravity	–	1.5	1.45	1.4
Tensile strength	MPa	110	100	85
Bend strength	MPa	150	130	110
Modulus in bend	GPa	7	6	5

*Courtesy of Scott Bader Co. Ltd.

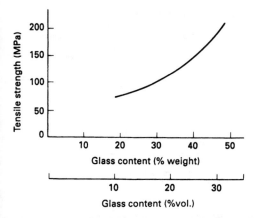

Figure 16.13 *A plot of tensile strength vs. glass content for CSM (Courtesy of Scott Bader Co. Ltd)*

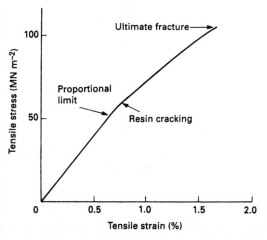

Figure 16.14 *A plot of tensile strength vs. tensile strain for CSM*

16.24 Thermosetting resins

The term 'synthetic resin' is used to describe man-made thermosetting pre-polymers because, although some are low-melting-point solids, many have the viscous, sticky consistency of naturally occurring resins, e.g. those secreted from coniferous trees.

Conventional terminology can be confusing as both the cross-linked polymers and the pre-polymers are commonly

Table 16.38 *Comparative thermal properties**

	Thermal conductivity (Wm⁻¹ K⁻¹)	Thermal expansion coefficient (×10⁻⁶ °C⁻¹)	Maximum working temperature (°C)
CSM/polyester GRP	0.2	30	175†
Unidirectional GRP	0.3	10‡	250†
Mild steel	50	12	400
Light alloy	200	23	200

*Courtesy of Scott Bader Co. Ltd.
†Depending on the type of polyester resin used and the application.
‡Measured longitudinally.

Table 16.39 *Relative rigidity of GRP sandwich laminates in bending**

Total thickness (mm)	Single GRP skin	Two GRP skins each 1.5 mm thick with centre core	Two GRP skins each 3 mm thick with centre core
1.5	0.018		
3	0.2	0.2	
6	1.6	1.3	1.6
10	5.3	3.7	5
13	12.5	7.3	10
19	42	18	29
25	100	34	58
32	195	52	94
38	337	77	143
44	536	104	197
50	800	145	266

*Courtesy of Fibreglass Ltd and Scott Bader Co. Ltd.

described as 'resins'. Thus, one component of a two-pack epoxy adhesive is an epoxy resin which, on reaction with the second component (the hardener or curing agent), gives a cured adhesive which is also frequently referred to as an epoxy resin. Strictly, the term 'resin' should be applied only to the precursor—the cured material is simply an 'epoxy'. This dual use is widespread but, fortunately, the specific context usually conveys the exact meaning.

Thermosetting resins, whether viscous liquids or soluble, fusible solids, generally consist of low- to intermediate-molecular-weight polymers. In both physical forms, they have specific reactive, functional groups either within or at the ends of the molecular chains.

The following generic chemical types are classified as thermosetting resins:

(1) Epoxies, furanes, polyurethanes and silicones
(2) Unsaturated polyesters, vinyl esters and certain acrylates
(3) Phenol, urea, resorcinol and melamine-formaldehydes.

An indication of the use of such resins in construction is shown in *Table 16.40*. Each resin type has advantages and disadvantages. These features are illustrated in *Table 16.41*. The cure rates or gel times for polyester resins vary not only with temperature but depend on the use of inhibitors, catalysts and accelerators. Some guidance is given on gel times in *Table 16.42*.

16.24.1 Properties
Some properties of thermosetting resins and comparisons with those of other materials are indicated in *Table 16.43*.

16.25 Silanes and siloxanes

Impregnating waterproof agents made of silanes/siloxanes (the terms are virtually synonymous) were developed relatively recently. They are quite effective as waterproof agents for concrete and inorganic construction materials. Increasing attention will be paid to this type of waterproofing agent which may improve the durability of concrete by means of suppressing alkali–silica reaction and salt attack.

The basic structure of silanes used for impregnating waterproof agents is a modified alkylalkoxysilane.

Factors such as water, oxygen, carbon dioxide, chloride ions, acids, ultraviolet radiation and temperature are involved in the deterioration of concrete. Of these, water plays a major role in the alkali–aggregate reaction, frost damage and salt attack (due to chloride ions). Therefore, it seems likely that the durability of concrete structures will be improved by waterproofing measures.

Waterproof agents for concrete are usually divided into two types: (1) agents which coat the entire surface of concrete; and (2) agents which penetrate into concrete and form a barrier within concrete. The former are called film-forming waterproof agents. Epoxy resins are representative of this type. The latter are called impregnating waterproof agents. *Figure 16.15* shows the commercially available impregnating waterproof agents.

Impregnating waterproof agents are made of organic solvents or emulsions. When applied to concrete, they penetrate deeply into the material, through the capillary pores, and give concrete a waterproof property. In this way, impregnating waterproof agents can prevent the invasion of water or ions (dissolved in water) into concrete. Impregnating waterproof agents can be subdivided into silicon-containing and silicon-free agents.

At present, more than 40 waterproof agents of the impregnating type are commercially available. The first waterproof agent of this type was made of siliconate, which was developed in 1965. Thereafter, products made of silicon resins were

Table 16.40 Uses of resin systems in construction and civil engineering

	Epoxy	Unsaturated polyester	Vinyl ester	Thermosetting acrylates	Furanes	Polyurethanes
Sealants for movement joints	(✓)*	–	–	–	–	✓†
Coatings/linings	✓	✓	✓	–	✓	✓
Chemically resistant renders	✓	✓	✓	✓	✓	✓
Impregnation	✓	✓	–	✓	–	✓
Adhesives/jointing	✓	✓	–	–	✓	✓
Floor toppings	✓	(✓)	(✓)	✓	–	✓
Repair materials	✓	✓	–	✓	–	(✓)
Gap filling/load bearing grouts	✓	✓	–	✓	–	–

*Little movement capability
†Usually in elastomeric form.

Table 16.41 *Relative advantages and disadvantages of resin-based systems*

Resin type	Benefits	Disadvantages
Epoxies	Excellent adhesion, good chemical resistance, little cure shrinkage, relatively tolerant of cold/wet cure conditions, good mechanical properties, wide scope in formulation, wide variation in pot-life and cure time, good shelf-life	High viscosity, health and safety, chalking caused by sunlight, slow cure, limited high-temperature performance, high thermal coefficient of expansion, critical mixing
Polyurethanes	Good chemical resistance, wide range of flexibility, good mechanical properties, little cure shrinkage, good adhesion, wide formulation scope	Health and safety, water sensitivity during cure, critical mixing, high thermal coefficient of expansion
Polyesters	Fast cure, low-temperature cure, mixing tolerance, chemical resistance, good against acids, good mechanical properties, low viscosity	High cure shrinkage, adhesion limited by shrinkage, health and safety, relatively short shelf-life, chemical resistance against alkalis can be poor, exotherm
Vinyl esters	Similar to polyesters but with improved chemical resistance and elevated-temperature performance	Similar to polyesters (but with improved chemical resistance)
Acrylates	Similar to polyesters but cure shrinkage has less consequence, good intercoat adhesion	Similar to polyesters
Furanes	Outstanding acid resistance and high-temperature performance	Sensitive to alkaline conditions/surfaces during cure, brittle, high coefficient of thermal expansion

Table 16.42   Gel times for polyester resins*

Polyester resin†	Approximate gel time	
	20°C	100°C
Without inhibitor	2 weeks	30 min
With inhibitor	1 year	5 h
With inhibitor and catalyst	1 week	5 min
With inhibitor, catalyst and accelerator	15 min	2 min

*Data from British Industrial Plastics Ltd.
†Inhibitor, 0.01% hydroquinone; catalyst, 1% benzoyl peroxide; accelerator, 0.5% dimethylaniline.

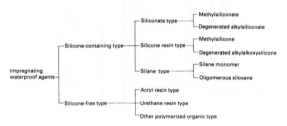

Figure 16.15   Specification of impregnating waterproof agents

developed in 1966, followed by products made of acryl resins and urethane resins (1974) and silanes (1978).

Of the silicones, methyl siliconate, i.e. silicone resin, is used for surface treatment of concrete, bricks and tiles. However, this substance has the following disadvantages:

(1) Insufficient penetration into the material
(2) Poor durability
(3) Poor resistance to alkalis
(4) Possibility of efflorescence if exposed to rain or sprayed water in 1–2 days after application.

Because of these properties, impregnating waterproof agents made of silicone resin have gradually been replaced by those made of silanes/siloxanes.

These products have gained wide acceptance in Japan but are less popular elsewhere. The relative effectiveness of a silane can be judged from *Figures 16.16* and *16.17*. The relationship between the moisture content of concrete and the depth of impregnation by a silane is shown in *Figure 16.18*. *Figure 16.19* indicates the expansion of a concrete wall in its untreated state compared with performance when coated with siloxane.

Table 16.43 A comparison of resin and ordinary Portland cement (OPC) based mortars and concretes

Property	Unit	Epoxy system	Polyester system	OPC system
Compressive strength	N mm^{-2}	55–110	55–110	15–70
Flexural strength	N mm^{-2}	20–50	25–30	2–5
Tensile strength	N mm^{-2}	10–20	10–20	1.5–4.5
Compressive modulus	kN mm^{-2}	0.5–20	2–10	20–30
Elongation at break	%	0–15	0–5	0
Linear coefficient of thermal expansion	×10^{-6} °C^{-1}	25–30	25–35	7–12
Water absorption, 7 days at 25°C	%	0–1	0.2–0.5	5–15
Maximum service temperature under load	°C	70–80	70–80	≥ 300
Rate of strength development at 20°C	hours or weeks	6–48 h	2–6 h	1–4 weeks

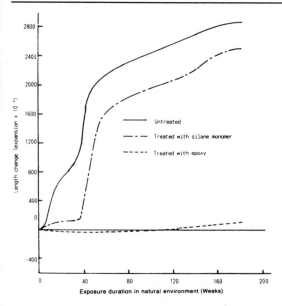

Figure 16.16 *Reduction in repellant ability of a silane monomer*

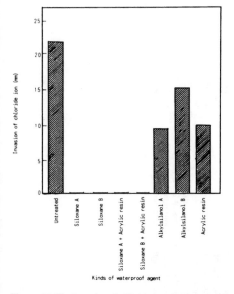

Figure 16.17 *Invasion of chloride ions into concrete treated with several waterproof agents. Specimens were immersed for 28 days in saturated sodium chloride solution after treatment*

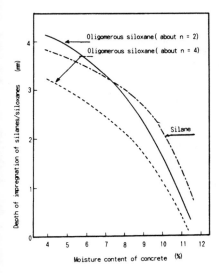

Figure 16.18 *Relationship between the moisture content of concrete and the depth of impregnation of silanes/siloxanes*

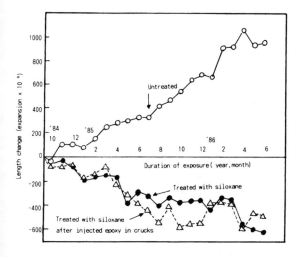

Figure 16.19 *Relationship between the duration of exposure and the expansion of a concrete wall. (○) Untreated; (●) treated with siloxane; (△) treated with siloxane after injection of epoxy resin into cracks*

Bibliography

DORAN, D.K. (ed.), *Construction Materials Reference Book*,
Butterworth–Heinemann, Oxford, (1992):
  Chapter 28: Ingold, T.S.
  Chapter 36: Tabor, L.J.
  Chapter 37: Coveney, V.A.
  Chapter 38: Grantham, M.G.
  Chapter 39: Dennis, R.
  Chapter 40: Allwood, W.J.
  Chapter 41: Crisfield, J., Cross, D.A. and Thompsett, D.J.
  Chapter 42: Bristow, R.G.
  Chapter 43: Arnold, G.H.
  Chapter 44: Burgoyne, C.J.
  Chapter 45: Andrews, R.H.
  Chapter 46: Hurley, S.A. and Humpage, M.L.
  Chapter 47: Nakano, K.

17 Slurries and grouts

Grouts and slurries are examples of special materials much used in civil engineering. The diversity of uses and properties of these systems is so great that it is difficult to give a concise overview. It is, however, appropriate to consider some of the fundamental similarities and differences.

Perhaps the most general common feature of grouts and slurries is that they tend to be complex non-Newtonian fluids (Newtonian fluids have no shear strength) and yet as many grouts and slurries are based on active swelling clays such as bentonite they tend to show somewhat similar fluid behaviour.

A distinction between grouts and slurries is that slurries are specifically used in slurry supported excavations such as dia- phragm walling, piling and slurry tunnelling. Grouts are gen- erally placed by injection and find a wider range of application from geotechnical engineering to structural engineering. A logi- cal distinction between grouts and slurries would be to refer to all non-setting systems as slurries and to reserve the term grout for setting materials. However, the term self-hardening slurry has been coined for cut-off wall slurries which are setting mate- rials. Thus grouts and slurries could be regarded as a spectrum of materials with setting grouts at one extreme and non-setting slurries at the other and self-hardening slurries as a link between the two systems. Though in practice the formulations for some self-hardening slurries may be apparently similar to those of geotechnical grouts.

An apparent similarity between grouts and slurries is that clays, Portland cements and the silicate solutions much used in grouting all involve silicon–oxygen chemistry (for example the silica layer of clays and the calcium silicate hydrates of cements). Despite this common chemistry the engineering beha- viour of the materials is quite distinct. However, the common fundamental chemistry does suggest some potential for chemi- cal compatibility of the materials and the potential for mixed systems. Indeed some of the most ancient and enduring con- struction materials were based on mixtures of clay and pozzo- lanic elements. In passing it may be noted that sodium aluminate is also used in chemical grouting and calcium alumi- nate hydrates are present in hydraulic cements.

For both grouts and slurries there is a wide range of special chemical systems not involving clays or hydraulic cements. In grouting these are now quite widely used though the relatively high cost tends to restrict their application to situations requir- ing their special properties. There are also chemical based slur- ries mainly developed from experience in the oil well drilling industry. These slurries are based on water soluble polymers and have been used in slurry tunnelling and to a rather more limited extent in diaphragm walling work.

Another link between slurries and grouts is that much of the equipment for testing fluid properties is common to both and is based on the equipment used for oil well drilling muds. Drilling muds have progressively developed from simple native clay systems to become complex chemical systems and it must be expected that there will be a similar trend with excavation slurries. Grouts are also used in oil well operations for cementing conductor pipes to the formation. The technology of well cementing has not significantly influenced geotechnical grouting or vice-versa but there are close parallels between oil well grouting and the filling of ducts in pre-stressed post tensioned concrete and offshore grouting for structural connections and repairs, etc.

It is not possible within the scope of this book to quote generalized properties for materials of such diversity.

Bibliography

JEFFERIS, S.A. In DORAN, D.K. (ed.), *Construction Materials Reference Book*, Butterworth–Heinemann, Oxford, Chapter 48 (1992)

18 *Stone*

18.1 Introduction

This chapter deals with stone as a construction material in its own right. It does not cover the use of the material as an aggregate for concrete, for which the reader is referred to Chapter 9. Stone has been used for building for centuries. Initially it was extracted from local quarries although there are early examples of the use of imported stone. Records suggest that Canterbury and Chichester cathedrals incorporated French limestone. Stone is used in a load-bearing mode, as a cladding and in other applications such as flow surfacing.

Building stone is derived from one of the following rock types:

- igneous
- metamorphic
- sedimentary.

Igneous rocks have crystallized from molten rock or 'magma'. The most prevalent UK igneous rock is granite, being a coarse-grained material containing at least 60% silica. The low porosity of such rocks makes them highly weather resistant.

Metamorphic rocks are those which have, due to heat and pressure, recrystallized from sedimentary deposits. Examples of this rock are slate and marble.

Sedimentary rocks (sandstones, limestones) are the principal UK building stones. They were formed in two stages:

(1) Deposited sediment
(2) Cementing of sediment under compaction and pressure to form a hard rock.

Sandstones have their origin in gneisses and igneous rock; they all contain silica. Sedimentary rocks will have been cemented by one of the following cements:

- siliceous (with silica)
- calcerous (with calcium carbonate)
- dolomite (with dolomite)
- ferruginous (with iron oxide)
- argillaceous (with clay).

The durability of the stone is affected by the type of cementing agent.

Limestones are usually composed of the shells and skeletons of marine creatures or from ooliths. They are all cemented with calcite; their durability is influenced more by the rock structure than its chemical nature.

Figure 18.1 gives an indication of the areas in the UK from which stone is quarried. Much stone is imported, as can be seen from the *Natural Stone Directory* published by Stone Industries.

18.2 Properties

Although stone has been used in a traditional manner for centuries it is now sometimes necessary to provide engineering justification for its use. Specialists from BRE and elsewhere have compiled a guide to properties which appear in *Table 18.1*. It should be emphasized, however, that individual stone properties may fall outside the stated ranges.

18.3 Deterioration

The critical factors in the deterioration of stone are the weathering agent and the physical/chemical composition of the stone. Prominent amongst the weathering agents are:

(1) Frost
(2) Soluble salts

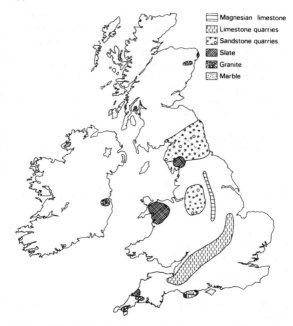

Figure 18.1 *Map to show main quarrying areas of building stone in the UK*

Table 18.1 Some properties of stone

Property	Units	Limestone	Sandstone	Marble	Granite	Slate
Compressive strength	MPa	14–255	34–248	69–241	96–310	138–206
Coefficient of linear thermal expansion	°C^{-1} (×10^6)	1.7–6.8	3.7–6.3	2.7–5.1	3.7–6.0	4.5–4.9
Moisture expansion	%	0.01	0.07			
Young's modulus	GPa	6.8–31.8	6.9–21	30–60	41.2–70.7	60–90
Density	g cm^{-3}	1.8–2.7	2.0–2.6	2.4–2.8	2.5–2.7	2.6–2.9
Porosity	%	≤ 50	5–31	0.6–2.3	0.4–2.3	0.1–4.3

(3) Acid deposits
(4) Moisture and temperature cycles.

18.3.1 Frost
When water freezes the process is accompanied by volume
increase. However, the mechanism by which damage occurs
has been found to be more complex than that due to water
expansion when ice forms. Pore size and distribution is now
thought to be a key factor in the damage process. The level
of water absorption/retention is determined by pore size. If
water retention is low, damage will not occur. Additionally
the thermodynamics of ice formation is also governed by pore
structure. Received wisdom suggests that single ice crystal
growth is largely responsible for frost damage.

18.3.2 Soluble salts
These cause deterioration in two ways:

(1) Crystallization similar to frost damage
(2) Moisture absorption producing volume expansion which
generates disruptive pressures.

18.3.3 Acid deposits
Acid gases such as sulphur dioxide attack stones containing
carbonate minerals such as calcite and dolomite. Two factors
are then at work:

(1) Erosion causing surface loss
(2) Crystallization damage.

18.3.4 Moisture and temperature cycles
Attack by moisture can be of a chemical or physical nature. If
the cementing material contains significant amounts of clay this
will expand on absorbing water thereby causing disruption.
Contour scaling is thought to be caused by a combination of
wetting and drying cycles combined with temperature variations. Such cycles may also cause chemical damage due to
pyrite oxidation.

18.4 Selection of stone

Whilst there are many codes and standards dealing with stone,
very little guidance is given on testing. The following are some
of the more significant standards:

BS 435: 1975 Specification for dressed natural stone kerbs,
channels, quadrants and setts
BS 680: Part 2: 1971 Specification for roofing slates
BS 743: 1970 Specification for materials for damp-proof
courses
BS 5390: 1976 (1984) Code of practice for stone masonry
BS 5534: Part 1: 1978 (1985), Part 2: 1986 Slating and tiling

BS 5642: Part 1: 1978, Part 2: 1983 Sills and copings
BS 6270: Part 1: 1982, Part 2: 1985 Code of practice for cleaning and surface repair of buildings
BS 8298: 1989 Code of practice for installation of natural stone cladding and lining.

18.5 Testing of stone

A variety of tests are required to assess the weathering characteristics of stone. The mechanisms by which stones decay varies. For example, limestones suffer from frost and salt attack; sandstones are generally resistant to frost but may suffer from salt attack and be susceptible to acid pollution. A guide to suitable tests for some stones is indicated in *Table 18.2*

In a crystallization test samples are exposed to cycles of soaking in a salt solution followed by oven drying. A measure of the durability of the stone is the degree of weight loss on drying; the higher the weight loss the lower the durability. The weight loss is used to allocate a durability class to the stone in accordance with the first two columns of *Table 18.3*. From this can be determined the suitability of a stone for a particular location in a building.

It is convenient to use the following sequence of building zones:

Zone 1   Paving, steps
Zone 2   Copings*, chimneys, cornices*, open parapets, finials, plinths*
Zone 3   Strings, plinths*, quoins, tracery hood moulds, solid parapets (excluding coping stones*), cornices*, mullions, sills
Zone 4   Plain walling

*A stone normally suitable for Zone 3 could be used for copings and cornices in Zone 2 if it were protected by lead. Similarly, a plinth in Zone 2 could be considered as Zone 3 if there were protection against rising damp.

The normal UK test is that developed by the BRE. The porosity of a stone is the volume of air in the stone expressed as a percentage of total stone volume. The saturation coefficient is the ratio of the volume of water absorbed by a sample immersed in cold water for 24 h to the volume of pore space in the sample.

The acid immersion test, applicable to sandstones and slates, is used to identify stone which is liable to decay when exposed to severe atmospheric pollution. *Table 18.4* gives an indication of acid strength for various materials and end uses. *Table 18.5* gives guidance on sampling.

Table 18.2 *Tests normally carried out for different types of stone*

Type of stone	End use	Crystallization test	Saturation coefficient	Acid immersion	Porosity	Wet/dry	Water absorption
Limestone	General	■	*		*		
Sandstone	General	*	*		*		
	Severe exposure	■		■			
Slate	Roofing			*		■	■
	Copings			■		■	
	Damp course			■		■	

■, These tests should always be carried out for the stone in question. *, These tests may be required for certain applications of the stone—see text for details. Note: the test conditions may vary for different stones; details are given in the main text.

Table 18.3 Effect of change of environment on the suitability for building zones 1–4

Limestone class	Crystallization loss (%)	Inland				Exposed coastal			
		Low pollution		High pollution		Low pollution		High pollution	
		No frost	Frost	No frost	Frost	No frost	Frost	No frost	Frost
A	<1	1–4	1–4	1–4	1–4	1–4	1–4	1–4	1–4
B	1–5	2–4	2–4	2–4	2–4	2–4	2–4	2*–4	2*–4
C	>5–15	2–4	2–4	3–4	3–4	3*–4	4		
D	>15–35	3–4	4	3–4	4				
E	>35	4	4	4*					
F	Shatters early in test	4	4						

*Probably limited to a 50-year life

Table 18.4 *Strength of acid for acid immersion tests*

Type of stone	Test or standard	Volume of 98% sulphuric acid (ml)	Volume of water (ml)
Sandstone	BRE 20%	300	2155
	BRE 40%	300	1015
Slate for roofing	BS 680	300	2100
Slate for sills and copings			
Type A	BS 5642	20	2370
Type B	BS 5642	1 vol. of acid prepared for type A slate in 4 volumes of water	

Table 18.5 *Number of samples, sample size and volume of acid required for the acid immersion test*

Type of stone	No. samples	Sample size (mm)	Volume of acid (ml)
Sandstone	6	$50 \times 50 \times 10$	200
Slate for roofing	3	$50 \times 50 \times t^*$	Totally immerse
Slate for sills and copings			
Type A	6	$25 \times 25 \times 25$	250
Type B	6	$25 \times 25 \times 25$	1000

*t, Thickness of slate.

18.6 Maintenance

Stone facades need periodic cleaning if good appearance is to be preserved. Dirt deposits include soot, wind-blown smoke, exhaust from vehicles (particularly from diesel engines). In the case of limestone these may chemically react with the stone to produce a black gypsum crust. This can usually be removed by water washing. Silica in sandstone is not water soluble and may require grit blasting or chemical cleaning.

Staining may be caused by organic growth, urban grime or metal run-off. Growths may be treated by toxic washes; metal stains are resistant to treatment although poulticing with ammonia solutions may be effective against copper staining (see BS 6270: 1982 for guidance). The treatment of iron staining varies with stone type (see *English Heritage Technical Handbook* for guidance).

Paint renewal may be difficult on porous stones. If water treatment is not satisfactory then paint removers to BS 3761: 1986 may be required. More ruthless methods such as paint strippers based on caustic soda, grinding or grit blasting are not recommended because of potential stone damage.

If surface damage is to be avoided then great care is required in the selection of cleaning method. The use of hydrofluoric acid on polished granite will, for example, totally destroy the highly polished finish.

18.7 Repair

Even well constructed stonework may need repair. Damage may have occurred due to explosion or long-term deterioration. One important consideration is the condition of mortar joints. If a facade or wall is showing distress then at least the pointing should be made good. The four principal methods of repair may be briefly summarized as follows:

- removal of loose surface material by de-scaling
- dressing back of stone to reveal sound surface
- stone replacement
- dressing back of stone then rebuilding to original profile in mortar (sometimes known as 'plastic repair').

Selection of repair method will usually depend on cost. Effective repair is a highly skilled job and should only be undertaken by experienced practitioners working to well-thought-out specifications. The reference to 'plastic repair' does not necessarily indicate the use of polymer-modified mortars but more to the consistency of the mix. Anyone responsible for the repair of stonework should be sceptical of the exaggerated claims made for some materials such as consolidants. The intention of these materials is to stabilize or retard the decay of stone—a claim rarely achieved in practice.

Acknowledgement

The assistance of senior staff at the Building Research Establishment, Watford, in producing the material for this chapter is particularly acknowledged.

Bibliography

BUTLIN, R.N. and ROSS, K.D. In DORAN, D.K. (ed.), *Construction Materials Reference Book*, Butterworth–Heinemann, Oxford, Chapter 49 (1992)

19 Timber

19.1 Introduction

Man has always had plenty of timber available for his needs. He has ample timber available today and will always have sufficient to meet his requirements. There will be changes in the availability of particular species and developments in the manufacture of different types of product, both in the form of solid timber and wood-based sheet materials, but, from an overall point of view, there will always be sufficient timber and timber products available for constructional use.

About one-third of the World's land surface is covered by forests, representing a growing stock of around 300 000 million m^3 of timber, of which nearly half are conifer trees which produce softwoods, the remainder being non-coniferous trees producing hardwoods. The pulp and paper industry takes around 40% of the value of the primary forest material with sawn wood amounting to about 38% and wood-based panel products amounting to about 22%. Set against the vast growing stock, the consumption of timber in the UK looks minute; in 1988 it was:

Sawn softwood	10 144 000 m^3
Sawn hardwood	1 410 000 m^3
Plywood	1 478 000 m^3
Chipboard	3 210 000 tonne.

Home production of timber in the UK amounts to about 21% of total timber consumption with about 16% of this being softwoods, 25% hardwoods, virtually no plywood and 45% other board materials such as chipboard. In relation to its weight and cost timber compares very favourably with many structural materials (see *Table 19.1*).

The bulk of the timber used for building in Europe is softwood rather than hardwood. The main reason for this preference is that in the past softwood has given the best combination of cost and performance and will tolerate a fair amount of abuse and yet still have an acceptable performance. Supply changes in the future mean that it will be necessary to use a wider range of species, both softwoods and hardwoods, with more variable properties than at present. Decorative hardwoods will undoubtedly achieve greater use as more leisure buildings are constructed with greater aesthetic appeal.

The large softwood-producing areas such as North America and Scandinavia now plan production against raw-material availability and are ensuring that supplies will be available as

Table 19.1 Comparison of structural performance of European redwood and other building materials

Material	Specific gravity (sg)	Tensile strength (T) (N mm^{-2})	T/sg	Modulus of elasticity (E) (N mm^{-2} × 10^3)	E/sg (×10^3)	Cost (£ tonne^{-1})
Redwood	0.5	92	184	7.7	15	200
Mild steel	7.84	470	60	207	26	400
Aluminium alloy	2.7	310	115	70	26	1100
Concrete	2.3	4	2	28.6	12.5	82

far ahead as one can see. Other developed countries are following this lead, for example New Zealand. Eventually, developing countries, with assistance, will follow and provide guaranteed supplies. Some of these may be species with which we are presently unfamiliar, but, by use of modern technology, promotion and marketing full utilization of them will be obtained.

'Softwood' and 'hardwood' are botanical terms and do not refer to the density or hardness of the wood. As it happens, most softwoods are fairly soft, though pitch pine and yew might be considered exceptions. In contrast, hardwoods can vary greatly in density and hardness; balsa is a hardwood as is obeche and so too are greenheart and *lignum vitae*.

Wood is composed of cells, which are made up of the cell wall, which varies in thickness in different timbers, and a central lumen or cavity. Most cells are long in relation to their width and are aligned axially along the tract, giving rise to the grain in wood. It is in terms of its function in the growth of the tree that some of the more important characteristics of wood can be understood. For example, one function of the stem of the tree, from which virtually all solid timber is cut, is to support the crown of leaves, the food-producing organ of the plant. The cells or individual fibres are aligned along the length of the stem so it can withstand the bending stresses induced by the movement of the crown in the wind. This gives the timber its relatively high longitudinal bending strength in relation to its weight and hence its use as, for example, floor joists in houses and other buildings.

An indication of the origins of UK supplies may be seen in *Figure 19.1*.

19.2 Properties

An extensive range of timber is available in the UK. The principal construction hardwoods and softwoods are shown in *Tables 19.2* and *19.3* respectively. Structural requirements are covered by BS 5268: Part 2: 1989. Eurocode 5 Part 1-1 *Timber—Common unified rules* is in preparation and will be issued as an ENV.

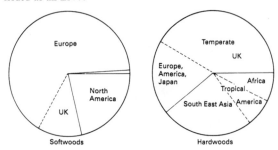

Figure 19.1 *Origins of UK timber supplies 1982*

Table 19.2 *Properties and uses of hardwoods*

Species	Colour	Density (kg m⁻³)	Texture	Moisture movement	Working qualities	Durability	Permeability	Uses
Abura *Mitragyna ciliata* W. Africa	Hardwood Light brown	580*	Medium/fine	Small	Medium	Non-durable	Moderately resistant	Interior joinery, mouldings
Afrormosia *Pericopsis elata* W. Africa	Hardwood Lightbrown	710	Medium/fine	Small	Medium	Very durable	Extremely resistant	Interior and exterior joinery. Furniture. Cladding
Afzelia/doussié *Afzelia* spp. W. Africa	Hardwood Reddish-brown	830*	Medium/coarse	Small	Medium/difficult	Very durable	Extremely resistant	Interior and exterior joinery. Cladding
Agba *Gossweilerodendron balsamiferum* W. Africa	Hardwood Yellowish-brown	510	Medium	Small	Good	Durable	Resistant	Interior and exterior joinery, trim. Cladding.
Andiroba *Carapa guianensis* S. America	Hardwood Pink to red-brown	640	Medium/coarse	Small	Medium	Moderately durable	Extremely resistant	Interior joinery
Ash, American *Fraxinus* spp. USA	Hardwood Grey, brown	670	Coarse	Medium	Medium	Non-durable	Permeable	Interior joinery, trim, tool handles

(*Continued*)

Table 19.2 *Continued*

Species	*Colour*	*Density ($kg\,m^{-3}$)*	*Texture*	*Moisture movement*	*Working qualities*	*Durability*	*Permeability*	*Uses*
Ash. European *Fraxinus excelsior* Europe	Hardwood White to light brown	710*	Medium/ coarse	Medium	Good	Perishable	Moderately resistant	Interior joinery.
Aspen *Populus tremuloides* Canada, USA	Hardwood Grey, white to pale brown	450	Fine	Large	Medium	Perishable/ non-durable	Extremely resistant	Interior joinery.
Balau† *Shorea* spp. S.E. Asia	Hardwood Yellow-brown to red-brown	980	Medium	Medium	Medium	Very durable	Extremely resistant	Heavy structural work, bridge and wharf construction
Balau, Red *Shorea* spp. S.E. Asia	Hardwood Purplish-red or dark red-brown	880	Medium	Medium	Medium	Moderately durable	Extremely resistant	Heavy structural work
Basralocus *Dicorynia guianensis* Surinam, French Guiana	Hardwood Lustrous brown	720	Medium	Medium/large	Medium	Very durable	Extremely resistant	Marine and heavy construction
Basswood *Tilia americana* N. America	Hardwood Creamy white to pale brown	420	Fine	Medium	Good	Non-durable	Permeable	Constructional veneer, turnery, piano keys, woodware

Beech, European *Fagus sylvatica* Europe	Hardwood Whitish to pale brown, pinkish-red when steamed	720	Fine	Large	Good	Perishable	Permeable	Furniture, interior joinery, flooring. Plywood.
Birch, American *Betula* spp. N. America	Hardwood Light to dark reddish-brown	710	Fine	Large	Good	Perishable	Moderately resistant	Furniture, plywood, flooring.
Birch, European *Betula pubescens* Europe, Scandinavia	Hardwood White to light brown	670	Fine	Large	Good	Perishable	Permeable	Plywood, furniture, tunery.
Cedar, Central/South American *Cedrela* spp. Central & S. America	Hardwood Pinkish-brown to dark reddish-brown	480	Coarse	Small	Good	Durable	Extremely resistant	Cabinet work, interior joinery.
Cherry, American *Prunus serotina* USA	Hardwood Reddish-brown to red	580	Fine	Medium	Good	Moderately durable	No information	Cabinet making, furniture. Interior joinery
Cherry, European *Prunus acium* Europe	Hardwood Pinkish-brown	630	Fine	Medium	Good	Moderately durable	No information	Cabinet making, furniture.

(Continued)

Table 19.2 Continued

Species	Colour	Density (kg m⁻³)	Texture	Moisture movement	Working qualities	Durability	Permeability	Uses
Chesnut, sweet *Castanea sativa* Europe	Hardwood Yellowish-brown	560	Medium	Large	Good	Durable	Extremely resistant	Interior and exterior joinery. Fencing
Danta *Nesogordonia papaverifera* W. Africa	Hardwood Reddish-brown	750	Fine	Medium	Good	Moderately durable	Resistant	Flooring, joinery, turnery
Ebony *Diospyros* spp. W. Africa, India, Sri Lanka	Hardwood Black, some grey/ black stripes	1030/1190	Fine	Medium	Medium	Very durable	Extremely resistant	Used primarily for decorative work. Turnery, inlaying
Ekki/azobé† *Lophira elata* W. Africa	Hardwood Dark red to dark brown	1070	Coarse	Large	Difficult	Very durable	Extremely resistant	Heavy construction, marine and freshwater construction. Bridges, sleepers, etc.
Freijo *Cordia goeldiana* S. America	Hardwood Golden brown	590	Medium	Medium/ small	Medium	Durable	No information	Furniture, interior and exterior joinery

	Type	Colour		Texture			Durability	Treatability	Uses
Gaboon *Aucoumea klaineana* W. Africa	Hardwood	Pinkish-brown	430	Medium	Medium	Medium	Non-durable	Resistant	Used principally for plywood and blockboard
Gedu nohor/edinam *Entandrophragma angolense* W. Africa	Hardwood	Reddish-brown	560	Medium	Small	Medium	Moderately durable	Extremely resistant	Furniture, exterior and interior joinery
Geronggang *Cratoxylon arborescens* S.E. Asia	Hardwood	Pink to red	550	Coarse	Medium	Medium	Non-durable	Permeable	Interior joinery
Greenheart† *Ocotea rodiaei* Guyana	Hardwood	Yellow/olive green to brown	1040	Fine	Medium	Difficult	Very durable	Extremely resistant	Heavy construction, marine and freshwater construction. Bridges, etc.
Guarea *Guarea cedrata* W. Africa	Hardwood	Pinkish-brown	590	Medium	Small	Medium	Very durable	Extremely resistant	Furniture, interior joinery, cabinet making

(*Continued*)

Table 19.2 Continued

Species	Colour	Density ($kg\,m^{-3}$)	Texture	Moisture movement	Working qualities	Durability	Permeability	Uses
Hickory *Carya* spp. N. America	Hardwood Brown to reddish-brown	830	Coarse	Large	Difficult	Non-durable	Moderately resistant	Striking tool handles, ladder rungs.
Idigbo *Terminalia ivorensis* W. Africa	Hardwood Yellow	560*	Medium	Small	Medium	Durable	Extremely resistant	Interior and exterior joinery, plywood
Iroko *Chlorophora excelsa* W. Africa	Hardwood Yellow-brown	660	Medium	Small	Medium/difficult	Very durable	Extremely resistant	Exterior and interior joinery, bench tops, constructional work
Jarrah† *Eucalpytus marginata* Australia	Hardwood Pink to dark red	820*	Medium	Medium	Difficult	Very durable	Extremely resistant	Heavy constructional work. Flooring
Karri† *Dryobalanops* spp. S.E. Asia	Hardwood Reddish-brown	770*	Medium	Medium	Medium	Very durable	Extremely resistant	Exterior joinery, decking, constructional use
Karri† *Eucalyptus diversicolor* Australia	Hardwood Reddish-brown	990	Medium	Large	Difficult	Durable	Extremely resistant	Heavy construction

Kauvula *Endospermum macrophyllum* Fiji	Hardwood Pale cream to straw-yellow	480	Medium to coarse	Small	Medium	Perishable	Permeable	Mouldings, interior joinery
Kempas† *Koompassia malaccensis* S.E. Asia	Hardwood Orange-red to red-brown	880	Coarse	Medium	Difficult	Durable	Resistant	Heavy constructional use
Keruing, apitong, gurjun, yang† *Dipterocarpus* spp. S.E. Asia	Hardwood Pinkish-brown to dark brown	740*	Medium	Large/medium	Difficult	Moderately durable	Resistant	Heavy and general construction. Decking.
Lauan see Meranti								
Limba/afara *Terminalia superba* W. Africa	Hardwood Pale yellow-brown/straw	560*	Medium	Small	Good	Non-durable	Moderately resistant	Furniture, interior joinery
Mahogany, Africa *Khaya* spp. W. Africa	Hardwood Reddish-brown	530	Medium	Small	Medium	Moderately durable	Extremely resistant	Furniture, cabinet work, joinery

(Continued)

Table 19.2 Continued

Species	Colour	Density (kg m⁻³)	Texture	Moisture movement	Working qualities	Durability	Permeability	Uses
Mahogany, American *Swietenia macrophylla* Central and S. America, especially Brazil	Hardwood Reddish-brown	560	Medium	Small	Good	Durable	Extremely resistant	Furniture, cabinet work, interior and exterior joinery.
Makoré *Tieghemella heckelii* W. Africa	Hardwood Pinkish-brown to dark red	640	Fine	Small	Medium	Very durable	Extremely resistant	Furniture, interior and exterior joinery. Plywood
Maple, rock *Acer saccharum* N. America	Hardwood Creamy white	740	Fine	Medium	Medium	Non-durable	Resistant	Excellent flooring timber. Furniture.
Maple, soft *Acer saccharinum* N. America	Hardwood Creamy white	650	Fine	Medium	Medium	Non-durable	Moderately resistant	Furniture. Interior joinery.
Mengkulang *Heritiera* spp. S.E. Asia	Hardwood Red, brown	720	Coarse	Small	Medium	Moderately durable	Resistant	Interior joinery. Construction. Plywood
Meranti, dark red/ dark red seraya/ red lauan *Shorea* spp. S.E. Asia	Hardwood Medium to dark red-brown	710*	Medium	Small	Medium	Variable, generally moderately durable to durable	Resistant to extremely resistant	Interior and exterior joinery. Plywood

Density is given in $kg\,m^{-3}$.

Name	Type / Colour	Density	Texture			Durability		Uses
Meranti, light red/ light red seraya/ white lauan *Shorea* spp. S.E. Asia	Hardwood Pale pink to mid-red	550*	Medium	Small	Medium	Variable, generally non-durable to moderately durable	Extremely resistant	Interior joinery. Plywood
Meranti, yellow/ yellow seraya *Shorea* spp. S.E. Asia	Hardwood Yellow-brown	660*	Medium	Small	Medium	Variable, generally non-durable to moderately durable	Extremely resistant	Interior joinery. Plywood
Merbau† *Intsia* spp. S.E. Asia	Hardwood Medium to dark red-brown	830	Coarse	Small	Moderate	Durable	Extremely resistant	Joinery, flooring, structural work
Nemesu *Shorea pauciflora* Malaysia	Hardwood Red-brown to dark red	710	Medium	Small	Medium	Moderately durable to durable	Resistant to extremely resistant	Interior and exterior joinery. Plywood
Niangon *Tarrietia utilis* W. Africa	Hardwood Reddish-brown	640*	Medium	Medium	Good	Moderately durable	Extremely resistant	Interior and exterior joinery. Furniture

(Continued)

Table 19.2 Continued

Species	Colour	Density (kg m⁻³)	Texture	Moisture movement	Working qualities	Durability	Permeability	Uses
Nyatoh *Palaquium* spp. S.E. Asia	Hardwood Pale pink to red-brown	720	Fine	Medium	Medium	Non-durable to moderately durable	Extremely resistant	Interior joinery, furniture
Oak, American red *Quercus* spp. N. America	Hardwood Yellowish-brown with red tinge	790	Medium	Medium	Medium	Non-durable	Moderately resistant	Furniture. Interior joinery
Oak, American white *Quercus* spp. N. America	Hardwood Pale yellow to mid-brown	770	Medium	Medium	Medium	Durable	Extremely resistant	Furniture, cabinet work.
Oak, European *Quercus robur* Europe	Hardwood Yellowish-brown	670/720	Medium/coarse	Medium	Medium/difficult	Durable	Extremely resistant	Furniture. Interior and exterior joinery. Flooring. Fencing
Oak, Japanese *Quercus mongolica* Japan	Hardwood Pale yellow	670	Medium	Medium	Medium	Moderately durable	Extremely resistant	Furniture. Interior joinery
Oak, Tasmanian *Eucalyptus delegatensis Eucalyptus obliqua Eucalyptus regnans* Australia, Tasmania	Hardwood Pale pink to brown	610/710	Coarse	Medium	Medium	Moderately durable	Resistant	Furniture, interior joinery

Species	Type / Colour	Density	Texture			Durability	Resistance	Uses
Obeche *Triplochiton scleroxylon* W. Africa	Hardwood White to pale yellow	390	Medium	Small	Good	Non-durable	Resistant	Interior joinery, furniture. Plywood
Opepe† *Nauclea diderrichii* W. Africa	Hardwood Yellow to orange-yellow	750	Coarse	Small	Medium	Very durable	Moderately resistant	Heavy constructional work. Marine and freshwater use. Exterior joinery. Flooring
Padauk *Pterocarpus* spp. W. Africa, Andamans, Burma	Hardwood Red to dark purple-brown	640/*850	Coarse	Small	Medium	Very durable	Moderately resistant to resistant	Interior and exterior joinery. Flooring
Pau marfim *Balfourodendron riedelianum* S. America	Hardwood Yellow	800	Medium	Large	Good	Non-durable	No information	Interior joinery, furniture. Flooring
Plane, European *Platanus hybrida* Europe	Hardwood Mottled red-brown	640	Fine	No information	Medium	Perishable	No information	Decorative purposes. Inlay work

(Continued)

Table 19.2   Continued

Species	Colour	Density (kg m⁻³)	Texture	Moisture movement	Working qualities	Durability	Permeability	Uses
Poplar *Populus* spp. Europe	Hardwood Grey, white to pale brown	450	Fine/medium	Large	Medium	Perishable non-durable	Extremely resistant	Box boards, turnery. Wood wool
Purpleheart *Peltogyne* spp. Central & S. America	Hardwood Purple to purplish-brown	880	Medium	Small	Medium/ difficult	Very durable	Extremely resistant	Heavy construction. Flooring.
Ramin *Gonystylus* spp. S.E. Asia	Hardwood White to pale yellow	670	Medium	Large	Medium	Non-durable	Permeable	Mouldings, furniture
Rosewood *Dalbergia* spp. S. America, India	Hardwood Medium to dark purplish-brown with black streaks	870*	Medium	Small	Medium	Very durable	Extremely resistant	Interior joinery, cabinet work, turnery
Sapele *Entandrophragma cylindricum* W. Africa	Hardwood Medium reddish-brown with marked stripe figure	640	Medium	Medium	Medium	Moderately durable	Resistant	Interior joinery, furniture, flooring
Sepetir *Sindora* spp. S.E. Asia	Hardwood Golden brown	640/*830	Medium	Small	Difficult	Durable	Extremely resistant	Joinery, furniture

Seraya—see Meranti

Sycamore *Acer pseudoplatanus* Europe	Hardwood	White or yellowish-white	630	Fine	Medium	Medium	Good	Perishable	Permeable	Turnery. Joinery
Taun *Pometia pinnata* S.E. Asia	Hardwood	Pale pinkish-brown	750	Coarse	Medium	Medium	Medium	Moderately durable	Moderately resistant	Structural work, turnery, joinery, furniture
Teak[†] *Tectona grandis* Burma, Thailand	Hardwood	Golden brown, sometimes with dark markings	660	Medium	Small	Medium	Medium	Very durable	Extremely resistant	Furniture, interior and exterior joinery. Boat building
Utile *Entandrophragma utile* W. Africa	Hardwood	Reddish-brown	660	Medium	Medium	Medium	Medium	Durable	Extremely resistant	Interior and exterior joinery. Furniture and cabinet work
Virola/baboen *Virola* spp. *Dialyanthera* spp. S. America	Hardwood	Pale pinkish-brown	430/*670	Medium	Medium	Medium	Medium	Non-durable	Permeable	Carpentry, furniture, plywood, mouldings

(Continued)

Table 19.2 Continued

Species	Colour	Density (kg m⁻³)	Texture	Moisture movement	Working qualities	Durability	Permeability	Uses
Wallaba *Eperua falcata Eperua grandiflora* Guyana	Hardwood Dull reddish-brown	910	Coarse	Medium	Medium	Very durable	Extremely resistant	Transmission poles, flooring, decking, heavy construction
Walnut, Africa *Lovoa trichilioides* W. Africa	Hardwood Yellowish-brown, sometimes with dark streaks	560	Medium	Small	Medium	Moderately durable	Extremely resistant	Furniture, cabinet work. Interior and exterior joinery
Walnut, America *Juglans nigra* N. America	Hardwood Rich dark brown	660	Coarse	Small/ medium	Good	Very durable	Resistant	Furniture
Walnut, European *Juglans regia* Europe	Hardwood Grey-brown with dark streaks	670	Coarse	Medium	Good	Moderately durable	Resistant	Furniture
Wenge *Millettia laurentii Millettia stuhlmannii* Central & E. Africa	Hardwood Dark brown with fine black veining	880	Coarse	Small	Good	Durable	Extremely resistant	Interior and exterior joinery. Flooring, turnery

*Density can vary by 20% or more.
†Structural properties included in BS 5268: Part 2: 1988.

Table 19.3 *Properties and uses of softwoods*

Species	Colour	Density (kg m⁻³)	Texture	Moisture movement	Working qualities	Durability	Permeability	Uses
Cedar of Lebanon *Cedrus libani* Europe	Softwood Light brown	580	Medium	Medium/small	Good	Durable	Resistant	Joinery, garden furniture, gates
Douglas fir† *Pseudotsuga menziesii* N. America and UK	Softwood Light reddish-brown	530	Medium	Small	Good	Moderately durable	Resistant/ extremely resistant	Plywood, interior and exterior joinery, construction. Vats and tanks
Hemlock, Western† *Tsuga heterophylla* N. America	Softwood Pale brown	500	Fine	Small	Good	Non-durable	Resistant	Construction, joinery
Larch, Japanese† *Larix kaempferi* Europe	Softwood Reddish-brown	560	Fine	Small	Medium	Moderately durable	Resistant	Stakes, general construction
Parana pine† *Araucaria angustifolia* S. America	Softwood Golden-brown with bright red streaks	550	Fine	Medium	Good	Non-durable	Moderately resistant	Interior joinery. Plywood

(Continued)

Table 19.3 Continued

Species	Colour	Density (kg m⁻³)	Texture	Moisture movement	Working qualities	Durability	Permeability	Uses
Pine, Corsican[†] *Pinus nigra* Europe	Softwood Light yellowish-brown	510	Coarse	Small	Medium	Non-durable	Moderately resistant	Joinery, construction
Pine, pitch[†] *Pinus palustris Pinus elliottii* Southern, USA	Softwood Yellow-brown to red-brown	670	Medium	Medium	Medium	Moderately durable	Resistant	Interior and exterior joinery, heavy construction
Pine, radiata *Pinus radiata* S. Africa, Australia	Softwood Yellow to pale brown	480	Medium	Medium	Good	Non-durable	Permeable	Furniture, packaging
Pine, Scots[†] *Pinus sylvestris* UK	Softwood Pale yellowish-brown to red-brown	510	Coarse	Medium	Medium	Non-durable	Moderately resistant	Construction, joinery
Pine, Southern[†] A number of species including *Pinus palustris, Pinus elliottii, Pinus echinata, Pinus taeda* Southern USA	Softwood Pale yellow to light brown	560*	Medium	Medium	Medium	Non-durable	Moderately resistant	Construction, joinery. Plywood

Name / Species / Origin	Type / Colour	Density	Texture	Movement	Workability	Durability	Treatability	Uses
Redwood, European† *Pinus sylvestris* Scandinavia, USSR	Softwood Pale yellowish-brown to red-brown	510	Medium	Medium	Medium	Non-durable	Moderately resistant	Construction, joinery, furniture
Spruce, Canadian† *Picea* spp. Canada	Softwood White to pale yellow	400/*500	Medium	Small	Good	Non-durable	Resistant	Construction, joinery
Spruce, Sitka† *Picea sitchensis* UK	Softwood Pinkish-brown	450	Coarse	Small	Good	Non-durable	Resistant	Construction, packaging, pallets
Spruce, Western white† *Picea glauca* N. America	Softwood White to pale yellow/brown	400/*500	Medium	Small	Good	Non-durable	Resistant	Construction, joinery
Western red cedar† *Thuja plicata* N. America	Softwood Reddish brown	390	Coarse	Small	Good	Durable	Resistant	Shingles, exterior cladding
Whitewood, European† *Picea abies* and *Abies alba* Europe, Scandinavia, USSR	Softwood White to pale yellowish-brown	470	Medium	Medium	Good	Non-durable	Resistant	Interior joinery, construction, flooring

(Continued)

Table 19.3 *Continued*

Species	Colour	Density (kg m^{-3})	Texture	Moisture movement	Working qualities	Durability	Permeability	Uses
Yew *Taxus baccata* Europe	Softwood Orange-brown to purple-brown	670	Medium	Small/medium	Difficult	Durable	Resistant	Furniture. Interior joinery

*Density can vary by 20% or more.
†Structural properties included in BS 5268: Part 2: 1988.

Moisture content (m/c) is a property critical to durability. Initially wood may contain as much water as wood fibre; drying is therefore a prerequisite to use. The risk of fungal attack is greatly reduced at a m/c below 25%; most undercover timber is below 20%. Densities vary with species and m/c; quoted values are averages at 15% m/c. Each increment of 1% in m/c will add about 0.5% to weight.

Moisture movement (defined as small, medium or large) refers to dimensional changes when timber is subjected to atmospheric variations. For structural purposes the category is not usually significant but may be for a decorative finish.

Durability is usually expressed by one of five classes based on the average life of a 50 mm × 50 mm section of heartwood in ground contact. Timber not in contact with the ground will usually have a longer life. Classes are:

(1) Very durable   > 25 years
(2) Durable   15–25 years
(3) Moderately durable   10–15 years
(4) Non-durable   5–10 years
(5) Perishable   < 5 years.

Preservative treatment should be specified where the natural durability is insufficient. Guidance on preservatives is given in BS 5268: Part 5: 1989, BS 5589: 1989 and BS 1186: Part 1: 1986.

Permeability is a measure of the ease with which timbers can be penetrated by preservatives. Categories are:

(1) Extremely resistant —absorbs only a small amount
(2) Very resistant   —difficult to penetrate more than 3– 6 mm
(3) Moderately resistant—6–18 mm in 2–3 h
(4) Permeable   —absorbs without difficulty.

19.3 Availability

Sawn timbers may be in regular or limited supply depending on availability from producers and levels of demand. At 1988 prices the ex-yard costs of kiln-dried timber, for parcels larger than 1.5 m^3, vary from below £400 m^3 to above £650 m^{-3}. These prices are for hardwoods with structural softwoods normally near the lower end of that scale. Widths in excess of 225/ 300 mm may be difficult to obtain even at a premium. Lengths up to 5 m are available; longer lengths may be obtained at a premium or by resorting to finger jointing (see BS 5291: 1984). Many cross-sectional sizes are available and an indication of the ranges is shown in *Tables 19.4* and *19.5*. The sizes in the standard are basic sawn sizes at a m/c of 20%. There are also references to the Canadian Lumber Standard (CLS) graded to the National Lumber Grades Authority (NLGA). This is timber machined on four sides with rounded arrises not greater than 3 mm and a m/c of 19% (approx).

Table 19.4 *Basic sizes (cross-sectional sizes mm) of sawn softwood (partly from BS 4471: Part 1: 1978)*

Thickness \ Width	75	100	125	150	175	200	225	250	300
16	■	■	■	■					
19	■	■	■	■					
22	■	■	■	■					
25	■	■	■	■	■	■	■	■	■
32	■	■	■	■	■	■	■	■	■
36	■	■	■	■					
38	■	■	■	■	■	■	■		
44	■	■	■	■	■	■	■	■	■
47*	■	■	■	■	■	■	■	■	■
50	■	■	■	■	■	■	■	■	■
63	■	■	■	■	■	■			
75	■	■	■	■	■	■	■	■	
100	■		■			■		■	■
150			■					■	
200				■					
250								■	
300									■

*This range of widths for 47 mm thickness will usually be available in constructional-timber quality only. The smaller sizes contained within the dotted lines are normally, but not exclusively, of European origin. The larger sizes outside the dotted lines are normally, but not exclusively, of North and South American origin.

Timber can be subjected to two other processes: regularizing and planing. By regularizing the width of a sawn section is made uniform throughout its length and is usually only performed on the two faces where accuracy is important. For sections up to 150 mm it removes 3 mm ±1 mm from the processed face. This process is normally used for structural timbers and when used it is important to specify the expected finished dimension.

Table 19.5 *Sizes (mm) of surfaced Canadian timber (CLS sizes)*

mm Thickness	Width	
38	63	▭
38	89	▭
38	140	▭
38	184	▭
38	235	▭
38	285	▭

Moisture content 19%
Tolerances: minus, nil; plus, no limitation.

Table 19.6 *Reductions from basic sizes to finished sizes by planing of two opposed faces (partly from BS 4471: 1987)**

	Reduction from basic size (mm)			
	15–35	*>35–100*	*>100–150*	*>150*
Construction timber	3	3	5	6
Matching interlocking boards	4	4	6	6
Wood trim not specified in BS 584	5	7	7	9
Joinery and cabinet work	7	9	11	13

*Note: floorings and wood trim are covered by separate British Standards: flooring BS 1297: 1967, wood trim BS 584: 1987

Planing requires the section to be surfaced on at least two opposite faces and usually all four of a rectangular section. *Table 19.6* shows the allowable reductions in size as set out in BS 4771: 1987.

19.4 Structural use of timber

This is covered by BS 5268 which is published in several parts. These deal with the following topics:

Part 1   Limit state design, materials and workmanship
Part 2: 1989 Permissible stress design, materials and workmanship
Part 3: 1985 Trussed rafter roofs

Part 4: 1987   Fire resistance of timber structures
Part 5: 1989   Preservation treatments for constructional timber
Part 6: 1988   Timber-framed walls
Part 7: 1989   Calculation basis for span tables (joints for domestic use, flat roofs, ceilings, ceiling binders, rafters and purlins)

The use of correct m/c in structural timber is essential; limiting values are shown in *Table 19.7*.

In BS 5268 the approach to design strength has been simplified by the use of nine strength classes. These cover a range from the weakest, lowest grade softwood to the densest, highest grade hardwood. These grades are obtained by a process known as stress grading which can be a visual or a machine technique. *Table 19.8* indicates the variation of bending stresses and elastic moduli with strength class. *Table 19.9* shows a range of softwood species which satisfy grading requirements SC1–SC5; *Table 19.10* indicates similar information for hardwoods for grades SC5–SC9. Stress grading techniques should comply with BS 4978: 1988 (Softwoods) or BS 5756: 1985 (Hardwoods).

Table 19.7 *Moisture content of timber for categories of end use (partly from BS 5268: Part 2: 1988)*

	Moisture content (%)	
	Average in service	Max. at erection
External uses, fully exposed	≥ 18	–
Covered and generally unheated	18	24
Covered and generally heated	16	21
Internal in continuously heated building	14	21

Table 19.8 *Dry grade stresses and moduli of elasticity for strength classes*

Strength class	Bending ($N\,mm^{-2}$)	Modulus of elasticity (mean) ($N\,mm^{-2}$)
SC1	2.8	6 800
SC2	4.1	8 000
SC3	5.3	8 800
SC4	7.5	9 900
SC5	10.0	10 700
SC6	12.5	14 100
SC7	15.0	16 200
SC8	17.5	18 700
SC9	20.5	21 600

Table 19.9 Softwood species/grade combinations which satisfy the requirements for strength classes SC1–SC5

Species	Origin	Grading rules*	SC1	SC2	SC3	SC4	SC5
					Grades to satisfy strength class		
Corsican pine	UK	BS 4978		GS	M50	SS	M75
Douglas fir	UK	BS 4978		GS	M50, SS	M75	
Douglas fir-larch	Canada, USA	BS 4978	No. 3		GS	SS	
		J & P			No. 1, No. 2	Select	
		Machine			1450f-1.3E	1650f-1.5E	1650f-1.5E 1800f-1.6E 1950f-1.7E 2100f-1.8E
European spruce	UK	BS 4978	GS	M50, SS			M75
				Machine graded to strength class			
Hem-fir	Canada, USA	BS 4978	No. 3		GS, M50	SS	
		J & P			No. 1, No. 2	Select	
		Machine			1450f-1.3E	1650f-1.5E	1650f-1.5E 1800f-1.6E 1950f-1.7E 2100f-1.8E
Larch	UK	BS 4978			GS	SS	
Panama pine	Any	BS 4978			GS	SS	
Pitch pine	Caribbean				GS	SS	
Radiata pine	New Zealand	BS 4978	Machine graded to strength class				

Table 19.9 Continued

Species	Origin	Grading rules*	Grades to satisfy strength class				
			SC1	SC2	SC3	SC4	SC5
Redwood	Imported	BS 4978			GS, M50	SS	M75
Scots pine	UK	BS 4978			GS, M50	SS, M75	
Sitka spruce	UK	BS 4978	GS	M50, SS	Machine graded to strength class		
	Canada	BS 4978			SS		
		J & P			Select		
Southern pine	USA	BS 4978			GS	SS	
		J & P			No. 1, No. 2, No. 3	Select	
		Machine			1450f-1.3E	1650f-1.5E	1650f-1.5E 1800f-1.6E 1950f-1.7E 2100f-1.8E
Spruce–pine–fir	Canada	BS 4978			GS, M50	SS, M75	
		J & P	No. 3		No. 1, No. 2	Select	
		Machine			1450f-1.3E	1650f-1.5E	1650f-1.5E 1800f-1.6E 1950f-1.7E 2100f-1.8E

Species	Source	Standard				
Western red cedar	Any	BS 4978	GS		SS	
Western whitewoods	USA	BS 4978 J & P	GS No. 3	No. 1, No. 2	SS Select	
Whitewood	Imported	BS 4978		GS, M50	SS	M75

*Timber graded to BS 4978: 1988. Timber graded to Canadian NLGA or American joist and plank (J & P) grades. Note: timber graded to North American structural light framing and stud grades are included in BS 5268: Part 2: 1989. Timber graded to North American machine stress-rated grades.

The machine grades MGS and MSS can be substituted for GS and SS, respectively.

The S6, S8, MS6 and MS8 ECE grades may be substituted for GS, SS, MGS and MSS, respectively.

The BS 4978 grading rules apply to timber of a minimum size of 35 mm × 60 mm.

The classification of NLGA and NGRDL grades into strength classes applied to timber of a minimum size of 38 mm × 114 mm.

Joist and plank No. 3 grade should not be used for tension members.

North American machine grades apply to a minimum size of 38 mm × 63 mm.

North American machine-graded timber is assigned into different strength classes depending upon the section size. See BS 5268: Part 2: 1988 for details.

BS 5268:1988 includes restrictions on fastener loads for: SC5 timbers (except pitch pine and Southern pine); British grown Sitka spruce and European spruce; US Western whitewoods; hem-fir and spruce-pine-fir in strength classes other than SC1 and SC2. See BS 5268: Part 2: 1988 for details.

Table 19.10 *Tropical hardwoods which satisfy the requirements for strength classes SC5–SC9 (graded to the HS grade of BS 5756: 1980)**

Standard name	Strength class
Iroko Jarrah Teak	SC5
Merbau Opepe	SC6
Kari Keruing	SC7
Balau Ekki Kapur Kempas	SC8
Greenheart	SC9

*BS 5756: 1980 refers to BS 5450: 1977 (sizes for hardwoods and methods of measurement)

19.5 Processed timber

Timber is frequently processed into board material. Some of the principal types of boarding are discussed below.

19.5.1 Plywood
This is defined as 'a wood-based panel product consisting of an assembly of plies bonded together, some or all of which are wood'. Veneer plywood is that in which all the plies are made of veneers orientated with their plane parallel to the surface of the panel. These materials are covered by BS 6100: Section 4.3: 1985. Plywood is a versatile product which can perform well under hazardous conditions whilst retaining comparatively high strength/weight properties. The main adhesives used for bonding of piles are urea formaldehyde and phenol formaldehyde. A third type of adheisve used mainly in the Far East is urea-formaldehyde fortified with melamine. Commonly available types of plywood available in the UK are marine (BS 1088/ 4079: 1966) and structural (BS 5268). In addition there is available a large variety of utility and decorative plywood.

19.5.2 Blockboard/laminboard (core plywoods)
These are composite boards having a core made up of strips of wood each ≤30 mm wide, laid separately or glued or otherwise joined together to form a slab, to each face of which is glued one or more veneers with the direction of the grain of the core strips running at right angles to that of the adjacent veneers. These materials are not usually marketed to any strict standards but manufacturers often use BS 3444: 1972 as a production aid. Urea formaldehyde is the normal adhesive used for blockboard

which is usually available in three or five ply form. This form of boarding permits, where applicable, the use of low-grade core veneers.

19.5.3 Wood chipboard
This is a particleboard made exclusively of wood from small particles and a synthetic resin binder. It is available in thicknesses from 3 mm to 50 mm and may be of uniform consistency or of layered construction. It is covered by BS 5669: 1979. The following types are readily available: standard (Type I to BS 5669: 1979), flooring (Type II to BS 5669), moisture resistant (Type III to BS 5669) and moisture resistant/flooring (Type II/III to BS 5669).

19.5.4 Fibre building boards
These are wood-based products usually exceeding 1.5 mm in thickness made from fibres of ligno-cellulosic material with the primary bond derived from the felting of fibres and their inherent adhesive properties. They are covered by BS 1149: 1989 and some are termed 'hardboards'. Types readily available include: standard (Type S), a dense product (usually $> 800 \, kg \, m^{-3}$) with a smooth face and a mesh pattern on the reverse; tempered, a higher density board ($> 960 \, kg \, m^{-3}$) to BS 5268 Part 2 in thicknesses from 3.2 mm to 12.7 mm; medium (LM of density in range $350 \, kg \, m^{-3}$ to $356 \, kg \, m^{-3}$, thicknesses 6.4 to 12.7 mm; HM of density in range of $560 \, kg \, m^{-3}$ to $800 \, kg \, m^{-3}$, thicknesses 8 mm to 12 mm); insulating board (less than $240 \, kg \, m^{-3}$, thicknesses 10 mm to 25 mm); bitumen-impregnated insulating board (boards with up to 35% bitumen may be used as expansion joint fillers in concrete); additionally there is a wide range of predecorated, embossed, surface-laminated and flame-retardant treated boards.

19.5.5 Flake boards
These may be waferboards made up from wood wafers or flakes at least 32 mm in length or oriented-strand boards (OSB) made from wood strands orientated in predetermined directions to simulate some characteristics of three-ply plywood. A British Standard in OSB is under consideration but the material is currently covered by Canadian Standards CAN 3-0188. For certain applications such as walling, flooring and roofing Agrément Certificates are relevant.

19.5.6 Medium density fibreboard (MDF)
MDF is a recent development in dry process board technology and is a sheet material manufactured from fibres of ligno-cellulosic material felted together with the primary bond derived from a bonding agent. The material is defined in BS 6100: its density usually exceeds $600 \, kg \, m^{-3}$ and is available in thicknesses from 6 mm to 50 mm. It is covered by BS 1142: 1989.

Table 19.11 Popular uses of the major wood-based sheet materials* in building, construction and allied applications

Application	Plywoods			Chipboards			Fibre building boards			
	WBP bonded ply	Non-WBP bonded ply	Blockboard & laminboard	Type I	Type II	Type III & II/III	Hardboard	Medium board	Insulating board (softboard)	MDF
Building elements										
Sheathing	×					×	×	×	×†	
Flat roofing	×					×			×†	
Roof sarking	×					×			×†	
Cladding	×							×	×†	×
Floor underlay	×						×	×	×	
Floor surface (dry)	×				×					
Floor surface (moisture hazard)	×					×				
Linings—interior partitions and wall panels	×	×		×			×	×	×	
Linings—ceilings and roofs	×	×		×			×	×	×	
Structural components										
Composite beams	×						×			
Truss gussets	×						×			
Stressed skin floor and roof panels	×				×	×	×			
Joinery, etc.										
Fascias and soffits	×							×		
Staircase construction	×		×		×	×				×

Window joinery								×
Mouldings and architraves	×		×					×
Furniture and built-in fitments	×		×	×		×		×
Door construction	×		×	×		×		
Temporary works								
Concrete formwork	×				×	×		
Signs and hoardings	×				×	×	×	
Other								
Shopfitting, display and exhibition work	×		×	×		×	×	×

*MDF, Medium-density fibre board; wood-based panel
†Bitumen-impregnated insulating board

19.5.7 Wood cement particleboard

This is manufactured from small particles of wood bonded with Portland or magnesite cement. There is no British Standard to cover this material which usually has a density in the range $1000\,\mathrm{kg\,m^{-3}}$ to $1200\,\mathrm{kg\,m^{-3}}$.

The uses to which major wood-based sheet/board materials can be applied are shown in *Table 19.11*.

19.6 Timber fasteners

These can be categorized into mechanical fasteners and glued joints.

19.6.1 Mechanical fasteners

In BS 5268 Part 2 the basic loads for fasteners are placed in four categories related to strength class (see *Table 19.12*). Four main types of mechanical fastening are used. These are: nails and screws (to BS 1202: 1974); bolts (to BS 4190: 1967 and BS 4320: 1968); tooth plate connectors (to BS 1579: 1966) and split ring connectors (to BS 1579: 1960).

19.6.2 Glued joints

These should be constructed in accordance with BS 5268: Part 2. The manufacture of timber sections for gluing should accord with BS 6446: 1984 and excessively resinous timber should be avoided. Some authorities are cautious in allowing total reliance on gluing and may insist on augmenting such joints with additional mechanical provision. In general glued joints are limited to members which are 50 mm thick in solid timber and with plywood not greater than 29 mm thick.

Adhesives should be appropriate to the environment of the structure or component and should accord with BS 1204: 1979. *Table 19.13* gives guidance on the selection of adhesives. Finger jointing is a particular technique involving the use of glue in order to produce relatively small cross-section structural timber to required lengths. This technique is restricted to non-principal

Table 19.12 *The four categories of the basic loads of fasteners (partly from BS 5268: Part 2: 1989)*

Joint strength classes	General application
1 SC1 and SC1	Low-strength softwoods
2 SC3 and SC4	Most commercial softwoods
3 SC5	High-strength softwoods and some hardwoods
4 SC6–SC9	Most hardwoods

Note: There are important exceptions in that softwoods (except pitch and Southern pine) machine graded to M75 should use fastener loads tabulated for SC3 and SC4.

Fast-grown conifers such as British-grown spruce, radiata pine from New Zealand and certain whitewoods from North America should use fastener loads tabulated for SC1 and SC2.

Table 19.13 *Permissible adhesive types*

Exposure category and conditions	Examples of adhesives*
High hazard	
Full exposure to the weather, e.g. marine structures where the glue-line is exposed to the elements. (Glued structures other than glued laminated members are not recommended for use under this exposure condition)	RF
Buildings with warm and damp conditions where a moisture content of 0.18 is exceeded and where the glue-line temperature can exceed 50°C, e.g. laundries, swimming pools and unventilated roof spaces	PF PF/RF
Chemically polluted atmospheres, e.g. chemical works and dye works	
External single-leaf walls with protective cladding	
Low hazard	
Exterior structures protected from sun and rain, roofs of open sheds and porches	RF
Temporary structures such as concrete formwork	
Heated and ventilated buildings where the moisture content of the wood will not exceed 0.18 and where the temperature of the glue-line will remain below 50°C, e.g. interiors of houses, halls, churches and other buildings	PF PF/RF MF/UF UF

*RF, resorcinol formaldehyde (BS 1204: Part 1: 1979, type WBP); PF, phenol formaldehyde (BS 1204: Part 1: 1979, type WBP); PF/RF, phenol resorcinol formaldehyde (BS 1204: Part 1: 1979, type WBP); MF UF, melamine urea formaldehyde (BS 1204: Part 1: 1979, type BR); UF, urea formaldehyde and modified UF (BS 1204: Part 1: 1979, type MR).

members and in load-sharing systems. BS 5291: 1984 lays down recommendations for joint efficiency (rating must exceed 50%) and manufacturing requirements. In calculation it is normal to assume that a finger joint does not affect the modulus of elasticity of the timber.

19.7 Glue laminated timber

Structural members of large cross-section and long length can be fabricated from small cross-section boards by the process of glued lamination, often referred to as 'glulam'.

A glued-laminated member comprises small cross-section boards layered up and glued so that the grain is parallel, in contrast to the configuration in plywood where adjacent veneers usually have their face grain at right angles. Individual boards may be connected by scarf or finger joints to behave as continuous laminations.

Manufacturers will produce laminated members in any size, section and profile. Most companies employ design engineers who will undertake detailed calculations and advise on construction.

A more recent development has been the introduction of standard straight beams which are available 'off the shelf'. These are made in a fixed range of sizes; the merchant or manufacturer usually stocks only a selection from the range and should be consulted about available sizes.

As a general guide, straight beams can range from 15 mm × 150 mm upwards although the small sizes will be uncompetitive with solid timber. Large sections of around 500 mm × 2000 mm spanning about 27.5 m are not uncommon. Gluing dictates low m/c which together with the random lay of pieces gives such members better dimensional stability than solid timber. Design guidance is given in BS 5268: Part 2.

The main species used for glue laminated timber are Scandinavian whitewood or redwood, Western hemlock or Douglas fir. Hardwood is normally only used for special strength or appearance. For economy it is permitted to vary the grade of timber in the layers which make up a composite member. Casein and synthetic resins are normally used and these should comply with BS 1204: Part 1: 1979. The thickness of laminations normally varies between 12 mm and 50 mm with the thinner dimensions usually dictated by bending requirements in curved members. Quality control of a high standard is required to produce elements which are fit for their purpose. Of particular importance is the strength and integrity of the adhesives which should be tested in accordance with BS 4169: 1988.

19.8 Trussed rafters

These are used mainly in domestic construction and are produced in a range of spans and configurations. Small section structural timbers are joined at the truss node points either by glued plywood gussets or more commonly by metal 'gang-nail'

Table 19.14 *Species of timber*

Standard name	Origin
Whitewood Redwood	Europe
Hem–fir Douglas fir–larch Spruce–pine–fir	Canada
Southern pine Hem–fir Douglas fir–larch	USA
Scots pine Corsican pine	UK

plates. The design and engineering justification of these rafters is dealt with in BS 5268: Part 3: 1985 using engineering data from BS 5268: Part 2. Such units are frequently justified by using the results of load tests. The range of species normally used in this type of construction are shown in *Table 19.14*.

19.9 Joinery timber

This timber is used for a variety of purposes—for example in the manufacture of window frames, doors and staircases. The general standard for this material is BS 1186: Part 1 (1986) and Part 1 (1988) dealing with materials and workmanship respectively.

Bibliography

SUNLEY, J.G. In DORAN, D.K. (ed.), *Construction Materials Reference Book*, Butterworth–Heinemann, Oxford, Chapter 50 (1992)

20 *Vermiculite*

Vermiculite is a member of the phyllosilicate group of minerals which embraces the chlorites, hydromicas and clay minerals. It is the geological name given to a group of hydrated laminar minerals which are magnesium aluminium iron silicates resembling mica in appearance. When subjected to heat, vermiculite has the unusual property of exfoliating, or expanding, due to the interlaminar generation of steam. Exfoliation produces a lightweight concertina-shaped granule.

By exfoliation and/or other forms of processing, such as grinding, the bulk density, size and aspect ratio of vermiculite can be varied to suit its end-use. As it is inorganic, easy and safe to handle, sterile after exfoliation, and, chemically and thermally stable, vermiculite is increasingly used where health and safety are important. In addition, vermiculite, particularly in the exfoliated form, is odourless and mould resistant, as well as being non-combustible, insoluble in water and all organic solvents.

Vermiculite is a very versatile mineral with chemical and physical properties that can be exploited in a wide range of applications. In addition to its use in the construction industry, vermiculite is used in the horticultural, automotive, agricultural, refractory and foundry industries.

Vermiculite is generally mined by open-cast methods. Extraction and beneficiation methods differ according to the location of the mine and the nature of the vermiculite occurrence.

The construction industry consumes 90% of the vermiculite imported into the UK.

The specific properties of vermiculite that are exploited in the building industry are:

(1) Lightweight for density modification of boards, premixes and concrete
(2) Adhesion to a wide range of substrates when used in plasters and cementitious premixes. Exfoliated vermiculite gives the best possible mixture for plastering on concrete surfaces
(3) Fire resistance
(4) Thermal insulation
(5) Anti-condensation properties
(6) Resistance to cracking and stress absorption
(7) Ease of use
(8) Acoustical modification.

The applications of vermiculite in the UK building industry in 1989 are indicated by the following estimate of consumption

Loft insulation	6%
Fire resistant boards and panels	12%
Fire protection premixes	16%
Fire protection cladding	26%
Gypsum plaster premixes	31%
Loosefill/lightweight aggregate	4%
Gypsum boards	5%

Typical technical data for vermiculite from the Palabora deposit is indicated in *Tables 20.1*, *20.2* and *20.3*. Vermiculite concrete can be used as fire protection (in accordance with BS 476) and for this purpose guidance is given in *Table 20.4*. General guidance on conditions imposed on materials in elevated temperatures is indicated in *Figure 20.1*.

Table 20.1 *Typical technical data*

Grades coarse 5.6–16 mm; fine 0.180–0.710 mm
Absorption capacity ≈ 500 cm^3 water/litre of exfoliated vermiculite
Angle of repose (crude ore) 38°–27° according to grade
Aspect ratio 7 : 1 to 30 : 1 (crude ore); up to 200 : 1 (processed)
Cation exchange up to 600 milliequivalents per kilogram
Chemical analysis (see *Table 20.2*)
Loose bulk density (crude ore) 600 to 1100 kg cm^{-3} (see *Table 20.3* for exfoliated)
Melting point (begins) 1330°C
Mohs hardness 1 to 2
Specific gravity 2.5
Specific heat 1.08 kj kg^{-1} K^{-1}
Specific surface area 5.0 to 7.7 m^2 g^{-1}
Thermal conductivity (exfoliated) 0.062–0.065 W m^{-1} °C^{-1} (ambient)
pH 8–9.5 (depending on grade/processing)
Weight loss at 105°C:4.5 to 6.5%
           950°C:8.0 to 10.0%

Table 20.2 *Typical composition of vermiculute*

Major constituents	Range (%)	Minor constituents	Content (%)
SiO_2	38.0–40.0	CaO	2.0
MgO	24.5–27.0	CO_2	1.5
Al_2O_3	8.0– 9.5	TiO_2	0.75
Fe_2O_3	5.0– 6.0	F	0.6
K_2	4.0– 6.0	CrO_3	0.15
H_2O	8.0–10.0	P_2O_5	0.06
		Cl	<0.05

Table 20.3 *Typical loose bulk densities of exfoliated vermiculite that can be achieved with standard grades of Palabora vermiculite*

Grade	Nominal particle size (crude) (mm)	Exfoliated loose bulk density (kg m⁻³)
Micron	−1.000 to +0.180	104–160
Superfine	−1.000 to +0.355	95–144
Fine	−2.800 to +0.710	88–122
Medium	−4.000 to +1.400	72–88
Large	−8.000 to +2.800	64–80
Premium	−16.000 to +5.600	56–72

Table 20.4 *Typical physical properties of vermiculite concrete*

Vermiculite: cement (vol.)	Air dry density* (kg m⁻³)	Minimum 28 day cube strength (N mm⁻²)	Thermal conductivity† (W m⁻¹ °C⁻¹)	Drying shrinkage (%)
8 : 1	400	0.70	0.094	0.35
6 : 1	480	0.95	to	to
4 : 1	560	1.23	0.158	0.45

*Low density = 320–800 kg m⁻³.
†Thermal insulation: $k = 0.086$ to 0.234 W m⁻¹ °C⁻¹; 0.074 to 0.200 kcal m m⁻² h⁻¹ °C⁻¹. Note that thermal insulation is a function of bulk density.

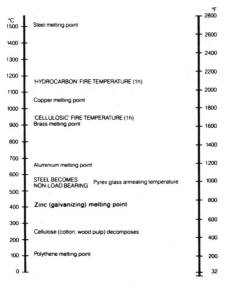

Figure 20.1 *Conditions imposed on a structural fire-protection material within petrochemical and off-shore marine environments*

Acknowledgement

The assistance of senior technical staff at Mandoval Ltd in producing the material for this chapter is particularly acknowledged.

Bibliography

CLARK, N.C. (ed.). In DORAN, D.K. (ed.), *Construction Materials Reference Book*, Butterworth–Heinemann, Oxford, Chapter 51 (1992)

Part Three: Appendices

Appendix A *Abbreviations, acronyms and organizations*

ACI	American Concrete Institute	
AF	Aluminium Federation	
ASTM	American Society for Testing Materials	
BACMI	British Aggregate and Construction Materials Industries	0171-730-8194
BBA	British Board of Agrement	01923-662900
BCA	British Cement Association	01344-762676
BCES	Building Centre Enquiry Service	01344-884
BCRA	British Ceramic Research Association	01782-45431
BCSA	British Constructional Steelwork Association	0171-839-8566
BDA	Brick Development Association	
BEC	Building Employers Confederation	0171-580-5588
BMP	Building Materials Producers	0171-222-5315
BPF	British Plastics Federation	0171-235-9483
BRE	Building Research Establishment	01923-664444
BS	British Steel	0171-388-5555
BSI	British Standards Institution	0171-629-9000
CIOB	Chartered Institute of Building	01344-243355
CIRIA	Construction Industry Research and Information Association	0171-222-8891
CITB	Construction Industry Training Board	01553-776677
CLS	Canadian Lumber Standard	
COSHH	Control of Substances Hazardous to Health	
CPDA	Clay Pipe Development Association	01494-791456
CS	Concrete Society	01753-662727
D of E	Department of Environment	0171-276-3000
DPC	Damp-proof course	
DTp	Department of Transport (Highways Agency)	0171-921-3881
ENV	European Pre-standard	
EH	English Heritage	0171-973-3000
FIP	Federation Internationale de la Precontrainte	0171-235-4535
GRCA	Glass Fibre Reinforced Cement Association	01952-811397
GGBFS	Ground granulated blast-furnace slag	
GRCA	Glass Fibre Cement Association	01942-825371

HAC	High-alumina cement	
HMSO	Her Majesty's Stationery Office	0171-873 9090
HRA	Hydraulics Research Association	
HSE	Health and Safety Executive	0171-221-0870
ICAS	International Annealed Copper Standard	
ICE	Institution of Civil Engineers	0171-222-7722
ISE	Institution of Structural Engineers	0171-235-4535
ISO	International Standards Organization	
MACEF	Mastic Asphalt Council Employers Federation	
MIG	Metal inert gas	
NAM	National Asphalt Mine-Owners and	
&MC	Manufacturing Council	
NBS	National Bureau of Standards (USA)	
NGRDL	National Grading Rules for Dimensioning Lumber (Canada)	
NHBC	National House-Building Council	01494-434477
NLGA	National Lumber Grades Authority (Canada)	
OPC	Ordinary Portland cement	
PFA	Pulverized fuel ash	
PIFA	Packaging and Industrial Film Association	
PVA	Polyvinyl acetate	
RH	Relative humidity	
RIBA	Royal Institute of British Architects	0171-580-5533
RICS	Royal Institute of Chartered Surveyors	0171-222-7000
SBR	Styrene–butadiene rubber	
SCI	Steel Construction Institute	01344-23345
SRC	Sulphate-resisting cement	
TIG	Tungsten inert gas	
TRADA	Timber Research and Development Association	01240-243091
TRL	Transport Research Laboratory	01347-773131
WAA	Water Authorities Association	0171-957-4567
WI	Welding Institute	01223-891162
WIMLAS	Wimpey Laboratories Assessment Service	0181-573-7744
ZDA	Zinc Development Association	0171-499-6636

NOTE: For further information see CIRIA Special Report 30, *Guide to Sources of Construction Information.*

Appendix B *The Greek alphabet*

Capital	Lower-case	Name	English transliteration
A	α	alpha	a
B	β	beta	b
Γ	γ	gamma	g
Δ	δ	delta	d
E	ϵ	epsilon	e
Z	ζ	zeta	z
H	η	eta	$\bar{e}$
Θ	θ	theta	th
I	ι	iota	i
K	κ	kappa	k
Λ	λ	lambda	l
M	μ	mu	m
N	ν	nu	n
Ξ	ξ	xi	x
O	o	omicron	o
Π	π	pi	p
P	ρ	rho	r
Σ	σ (ς at end of word)	sigma	s
T	τ	tau	t
Υ	υ	upsilon	u
Φ	ϕ	phi	ph
X	χ	chi	kh
Ψ	ψ	psi	ps
Ω	ω	omega	$\bar{o}$

Appendix C *Table of atomic symbols*

Element	Symbol	Element	Symbol
Actinium	Ac	Mercury	Hg
Aluminum	Al	Molybdenum	Mo
Americium	Am	Neodymium	Nd
Antimony	Sb	Neon	Ne
Argon	Ar	Neptunium	Np
Arsenic	As	Nickel	Ni
Astatine	At	Niobium	Nb
Barium	Ba	Nitrogen	N
Berkelium	Bk	Nobelium	No
Beryllium	Be	Osmium	Os
Bismuth	Bi	Oxygen	O
Boron	B	Palladium	Pd
Bromine	Br	Phosphorus	P
Cadmium	Cd	Platinum	Pt
Calcium	Ca	Plutonium	Pu
Californium	Cf	Polonium	Po
Carbon	C	Potassium	K
Cerium	Ce	Praseodymium	Pr
Cesium	Cs	Promethium	Pm
Chlorine	Cl	Protactinium	Pa
Chromium	Cr	Radium	Ra
Cobalt	Co	Radon	Rn
Copper	Cu	Rhenium	Re
Curium	Cm	Rhodium	Rh
Dysprosium	Dy	Rubidium	Rb
Einsteinium	Es	Ruthenium	Ru
Erbium	Er	Samarium	Sm
Europium	Eu	Scandium	Sc
Fermium	Fm	Selenium	Se
Fluorine	F	Silicon	Si
Francium	Fr	Silver	Ag
Gadolinium	Gd	Sodium	Na
Gallium	Ga	Strontium	Sr
Germanium	Ge	Sulphur	S
Gold	Au	Tantalum	Ta
Hafnium	Hf	Technetium	Te
Helium	He	Tellurium	Te
Holmium	Ho	Terbium	Tb
Hydrogen	H	Thallium	Tl
Indium	In	Thorium	Th
Iodine	I	Thulium	Tm
Iridium	Ir	Tin	Sn
Iron	Fe	Titanium	Ti
Krypton	Kr	Tungsten	W
Lanthanum	La	Uranium	U
Lead	Pb	Vanadium	V
Lithium	Li	Xenon	Xe
Lutetium	Lu	Ytterbium	Yb
Magnesium	Mg	Yttrium	Y
Manganese	Mn	Zinc	Zn
Mendelevium	Md	Zirconium	Zr

Appendix D *Public libraries (UK) holding sets of British Standards*

The following are UK public libraries which hold sets of British Standards. Attention is drawn to the law of copyright; no part of a BSI publication may be reproduced without the prior permission of BSI. Students and lecturers will often be able to find sets in their college libraries.

England

Avon	Bristol	*Central Library*
Bedfordshire	Bedford	*Public Library*
	Luton	*Central Library*
Berkshire	Bracknell	*Central Library*
	Reading	*Central Library*
	Slough	*Central Library*
Buckinghamshire	Aylesbury	*Public Library*
	Milton Keynes	*Central Library*
Cambridgeshire	Cambridge	*Central Library*
Cheshire	Chester	*Public Library*
	Crewe	*Public Library*
	Ellesmere Port	*Central Library*
	Warrington	*Public Library*
Cleveland	Hartlepool	*Central Library*
	Middlesbrough	*Central Library*
Cumbria	Barrow-in-Furness	*Central Library*
Derbyshire	Derby	*Central Library*
	Matlock	*Public Library*
Devonshire	Exeter	*Central Library*
	Plymouth	*Central Library*
Dorset	Poole	*Arndale Reference Library*
Durham	Durham	*Public Library*
Essex	Colchester	*Public Library*
	Grays	*Public Library*
	Southend-on-Sea	*Public Library*
	Witham	*Public Library*
Gloucestershire	Cheltenham	*Public Library*
	Gloucester	*Gloucestershire Technical Information Service*
Hampshire	Basingstoke	*Public Library*
	Farnborough	*Public Library*
	Portsmouth	*Central Library*
	Southampton	*Central Library*
	Winchester	*Public Library*
Hereford and Worcester	Redditch	*Public Library*
Hertfordshire	Stevenage	*Central Library*

Humberside	Grimsby	*Central Library*
	Hull	*Central Library*
	Scunthorpe	*Public Library*
Kent	Chatham	*Public Library*
Lancashire	Blackburn	*Public Library*
	Burnley	*Central Library*
	Lancaster	*Public Library*
	Preston	*Central Library*
	Skelmersdale	*Public Library*
Leicestershire	Leicester	*Information Centre*
Lincolnshire	Lincoln	*Public Library*
Greater London	Barking	*Central Library*
	Barnet	*Hendon Central Library*
	Bexley	*Central Library*
	Brent	*Central Library*
	Bromley	*Central Library*
	Croydon	*Central Library*
	Ealing	*Central Library*
	Enfield	*Palmers Green Public Library*
	Greenwich	*Woolwich Public Library*
	Hammersmith	*Central Library*
	Haringey	*Central Library*
	Harrow	*Central Reference Library*
	Havering	*Romford Central Library*
	Hounslow	*Feltham Public Library*
	Islington	*Central Library*
	Kensington and Chelsea	*Central Library*
	Lambeth	*Tate Library*
	Lewisham	*Deptford Public Library*
	Merton	*Wimbledon Public Library*
	Newham	*Stratford Public Library*
	Redbridge	*Ilford Central Library*
	Sutton	*Central Library*
	Waltham Forest	*Central Library*
	Wandsworth	*Battersea District Library*
	Westminster	*Central Reference Library*
Greater Manchester	Ashton-under-Lyne	*Public Library*
	Bolton	*Public Library*
	Bury	*Central Library*

	Manchester	*Central Library*
	Oldham	*Central Library*
	Rochdale	*Central Library*
	Stockport	*Central Library*
	Wigan	*Central Library*
Merseyside	Liverpool	*Central Reference Library*
	St Helens	*Central Library*
Norfolk	Norwich	*Central Library*
Northamptonshire	Northampton	*Central Library*
Northumberland	Morpeth	*Public Library*
Nottinghamshire	Mansfield	*Central Library*
	Nottingham	*Central Library*
Oxfordshire	Oxford	*Central Library*
Somerset	Bridgwater	*Public Library*
Staffordshire	Burton-upon-Trent	*Public Library*
	Cannock	*Public Library*
	Stafford	*Public Library*
	Stoke-on-Trent	*Hanley Central Library*
Suffolk	Lowestoft	*Central Library*
Surrey	Woking	*Public Library*
East Sussex	Brighton	*Reference Library*
West Sussex	Crawley	*Public Library*
Tyne and Wear	Gateshead	*Public Library*
	Newcastle-upon-Tyne	*Central Library*
	North Shields	*Central Library*
	South Shields	*Central Library*
	Washington	*Central Library*
Warwickshire	Rugby	*Business Information Service*
West Midlands	Birmingham	*Central Library*
	Coventry	*Public Library*
	Dudley	*Public Library*
	Walsall	*Central Library*
	West Bromwich	*Sandwell Central Library*
	Wolverhampton	*Central Library*
Wiltshire	Trowbridge	*Public Library*
South Yorkshire	Barnsley	*Central Library*
	Doncaster	*Central Library*
	Rotherham	*Central Library*
	Sheffield	*Central Library*
North Yorkshire	Northallerton	*Public Library*
	York	*Public Library*
West Yorkshire	Bradford	*Central Library*
	Dewsbury	*Public Library*
	Huddersfield	*Public Library*
	Leeds	*Central Library*
	Wakefield	*Central Library*

Wales

South Glamorgan	Cardiff	*Public Library*
West Glamorgan	Swansea	*Central Library*

Scotland

Central	Falkirk	*Public Library*
Grampian	Aberdeen	*Public Library*
Lothian	Edinburgh	*Central Library*
Strathclyde	East Kilbride	*Central Library*
	Glasgow	*Mitchell Library*
Tayside	Dundee	*Central Library*

Northern Ireland

Antrim	Ballymena	*Public Library*
	Belfast	*Central Library*
Connagh	Portadown	*Information Services*
Tyrone	Omagh	*Public Library*
		Headquarters

Index